Teaching Music Improvisation with Technology

Teaching Music Improvisation with Technology

Michael Fein

OXFORD
UNIVERSITY PRESS

OXFORD
UNIVERSITY PRESS

Oxford University Press is a department of the University of Oxford. It furthers the University's objective of excellence in research, scholarship, and education by publishing worldwide. Oxford is a registered trade mark of Oxford University Press in the UK and certain other countries.

Published in the United States of America by Oxford University Press
198 Madison Avenue, New York, NY 10016, United States of America.

Library of Congress Cataloging-in-Publication Data
Names: Fein, Michael, author.
Title: Teaching music improvisation with technology / Michael Fein.
Description: New York : Oxford University Press, [2017] | Includes bibliographical references and index.
Identifiers: LCCN 2016027884 (print) | LCCN 2016028834 (ebook) | ISBN 9780190628260 (pbk. : alk. paper) | ISBN 9780190628253 (cloth : alk. paper) | ISBN 9780190628277 (updf) | ISBN 9780190628284 (epub)
Subjects: LCSH: Improvisation (Music) | Jazz—Instruction and study. | Music—Computer programs. | Music and the Internet.
Classification: LCC MT68 .F45 2017 (print) | LCC MT68 (ebook) | DDC 781.3/6071—dc23
LC record available at https://lccn.loc.gov/2016027884

9 8 7 6 5 4 3 2 1

Paperback printed by Webcom, Inc., Canada
Hardback printed by Bridgeport National Bindery, Inc., United States of America

To my wife, Aimee, and wonderful children, Julian, Grayson, and Holden:

I wouldn't have been able to work through this project without your love and support. You inspire me and keep me laughing throughout life.

To my parents:

Thank you for attending every concert, applauding past when the rest of the audience stopped, and for encouraging comments like "personal best."

Preface

I started playing saxophone in fourth grade and I haven't stopped since. My interest in saxophone coincided with a great leap in the availability and accessibility of music technology throughout the 1990s and 2000s. Whether it was using a minidisc recorder (anyone remember those?) to record my lesson or slowing down a recording with digital audio software to help transcribe a difficult section of a Dexter Gordon solo, technology has consistently been a valuable tool to assist my practicing and development as an improviser. Since 2003, I have taught jazz ensemble and music technology electives to high school students full-time in Havertown, Pennsylvania. The jazz ensemble students are typically some of the best trained musicians in the school while tech elective students typically have limited prior music experience and in many cases don't read traditional music notation or perform in the band, orchestra, or choir. Interestingly, I have found that technology allows me to meet the students at their current level and take them further musically than I could have ever done without it. All of my lessons and activities *start* with a musical goal and *end* with a technology application. It is true that there is no replacement for hours of practicing, and technology alone will not turn you into a brilliant improviser. This book illustrates how technology can be used to *support* improvisational growth and musical development every step of the way in both the practice room and the music classroom.

In this book, I guide you, the reader, through the basic mechanics of improvising and the essential music theory elements needed to improvise and teach improvisation, including modal improvisation, the blues, ii-V-I progressions, and chord bracketing. Each chapter focuses on a different area of music technology and provides concrete hands-on activities to put theory into practice. I also make iPad connections throughout the book, so each chapter has relevance regardless of whether the reader is working on a desktop, laptop, or tablet.

Acknowledgments

I have been blessed with exceptional teachers throughout my life. The guidance of these amazing educators pushed me to develop into the musician and educator that I have become today. As I look back on my musical development, I think I had the right teachers at the right time and I am so appreciative of everything they did for me.

- Roy Martin—Roy was my private saxophone teacher from fourth through twelfth grade. He set me up with solid saxophone fundamentals and pushed me throughout my development in both the jazz and classical worlds.
- Thomas Rudolph—Tom was my middle school concert band and jazz ensemble director. He really got me into the world of music technology and he is still a mentor I frequently lean on for advice.
- Gerry DeLoach—Gerry was my high school jazz ensemble director. He is truly a great player and teacher. He made improvising approachable and I still remember him taking time during his lunch to help me navigate my way through tunes like "Take the A Train."
- Ralph Bowen—Ralph was my private saxophone teacher and big band director at Rutgers University where I received my bachelor's degree. As a high school senior, I saw Ralph perform at Ortlieb's Jazz Haus in Philly, and I knew I wanted to study with him and attend Rutgers. Ralph's scale and chord exercises helped me develop the technical skills I needed to improvise better, and he provided a solid framework for developing improvisational skills in a sensical manner.
- Denis DiBlasio—Denis was my private saxophone teacher at Rowan University where I received my master's degree. Denis is one of the warmest people I've known. He took my playing to the next level by teaching me arranging concepts, having me memorize lots of melodies and solo transcriptions, and encouraging me to let my ear guide my improvisation.

I can only hope that I can have a shred of the impact on my students that these educators had on me.

Contents

About the Companion Website

www.oup.com/us/teachingmusicimprovisationwithtechnology

Oxford has created a website to accompany *Teaching Music Improvisation with Technology*. Material that cannot be made available in a book, namely, audio examples of various activities presented in the book, is provided here. The reader is encouraged to consult this resource in conjunction with each hands-on activity throughout each chapter. Examples available online are indicated in the text with Oxford's symbol [▶].

1 INTRODUCTION TO THE MECHANICS OF IMPROVISATION

This chapter first focuses on establishing a clear definition of what improvisation is (and is not), provides a framework for developing improvisation skills, and discusses basic guidelines for selecting appropriate repertoire for beginning improvisation. Next, it provides an overview of technology hardware and software appropriate for teaching improvisation in a private lesson, classroom, or lab setting. I also survey existing materials including the Aebersold, Essential Elements, and Standard of Excellence methods and SmartMusic. SmartMusic is a subscription-based application that includes the entire Essential Elements and Standard of Excellence methods. SmartMusic also includes outstanding ear training activities, blues/ii-V-I/iii-VI-ii-V-I scale and arpeggio exercises, rhythm section accompaniments, and transposition exercises. A user can even change the tempo and key of an existing audio recording! In the iPad connection portion of this chapter, I discuss available exercises and materials in the free SmartMusic application that comes with your paid SmartMusic subscription. Finally, I wrap up the chapter with a hands-on activity to "break the ice" and get students feeling comfortable and free to improvise.

Improvisation is . . .

[▶] *See Example 1.1 on the companion website for a PDF of my "Improvisation Is . . ." Keynote Presentation.*

. . . composing in the moment with restrictions

These restrictions are generally the rhythm and harmony of the tune. The improviser selects rhythms and notes that cause tension and release as he or she plays. Improvised notes that do not fit in the chord cause tension until they release into chord tones. The improviser chooses rhythms that groove with the rhythm and those that cause syncopation. So, when the improviser begins performing, he or she isn't really dealing with a blank canvas; rather, the improviser begins with a backdrop (harmony and rhythm of the tune) and adds the foreground. Although the improviser is free to play anything, whatever he or she creates will interact with the harmony and rhythm, and all of these decisions are being made instantaneously throughout the improvised performance.

Here's a brief analogy:

My son recently received a Star Wars sticker book for the holidays. In the front of the book, there are various movie scenes such as a desert, space, and what looks to be the inside of a spaceship. In the back of the book are various stickers that can be applied to each scene. Grayson is able to tell a story by building a sticker scene based on the background that already exists on the page. My older son, Julian, and I always get a chuckle from an unexpected sticker such as one that sits sideways or obviously doesn't fit (I mean we all know Wampas like the cold and don't live in the desert!). So as Grayson improvises (creates on the spot), he interacts with the background to create tension/release and groove/syncopation.

The great composer, Igor Stravinsky, commented on composing music with restrictions in the book *Poetics of Music in the Form of Six Lessons*:

> What delivers me from the anguish into which an unrestricted freedom plunges me is the fact that I am always able to turn immediately to the concrete things that are here in question. . . . [I]n art as in everything else, one can build only upon a resisting foundation: whatever constantly gives way to pressure, constantly renders movement impossible. My freedom thus consists in my moving about within the narrow frame that I have assigned myself for each one of my undertakings. The more constraints one imposes, the more one frees one's self.[1]

Stravinsky is certainly not a composer we think of as bland or uncreative; I'd classify him as one of the most daring composers of the early 20th century, always pushing the boundaries of musical acceptability. Throughout this text I will recommend improvising within a narrow framework such as a specific scale or set of chord tones. Students must improvise in the box with significant restrictions to maximize creativity.

. . . limited by three main factors: technical ability, ear development, and music theory knowledge.

1. **Technical ability:** My main instrument is saxophone. I have spent countless hours over the past 25 years practicing the saxophone, learning my scales, arpeggios, melodies, etc., on *that* instrument. When I improvise on saxophone, I have more technical freedom because the instrument feels almost like an extension of myself. In fact, when I'm examining a new piece of music, I often find myself fingering notes on my "air sax" to more accurately hear the notes in my head. I also play clarinet, flute, keyboard, and guitar. Guitar is my weakest instrument by far and because of my drastic technical limitations on guitar, I can use only simpler phrases when I improvise on that instrument. I find the same applies to my younger students. The more they become familiar with their instrument by practicing all of their fundamental technique, the more freedom they have when improvising. Encourage students to focus on developing proper fundamental technique on their instrument including tone, rhythmic feel, articulation, and technical facility.
2. **Ear development:** The best improvisers hear music in their head and produce it immediately and effortlessly on their instrument. It is essential to rigorously

1 Igor Stravinsky, *Poetics of Music in the Form of Six Lessons* (Cambridge, MA: Harvard University Press, 1970), 65–66.

train your ear to identify intervals, practice sight-singing melodies (both tonal and atonal), and perform familiar melodies in multiple keys. I encourage my students to select a simple melody (such as *Mary Had a Little Lamb, Happy Birthday*, etc.) and practice playing it in all 12 keys (or at least a handful of keys). I also find it helpful to perform melodies by ear such as those from pop tunes on the radio; instead of singing the melody, perform it on an instrument. It is also essential to listen to a lot of music, especially by great improvisers. Focus on the artist's phrases, tone, and articulation. Many of my jazz students will purchase a book of jazz solo transcriptions and try to perform them without listening to the original recording. This is exactly the wrong thing to do. You first need to listen to the recording until you can just about sing the entire thing. Then use the solo transcription as an aid to perform the solo exactly as the artist did with the exact same inflections and tone. Through this listening and imitation, the student will become a better improviser simply by osmosis.

3. **Music theory knowledge:** First, students must have a basic grasp of music theory including key signatures, 12 major scales, 12 harmonic minor scales, and the 7 modes of the major scale. Second, a student must be able to execute that knowledge on his or her instrument. Of course, this is a tall order for many high school or college-level students and far beyond the reach of most elementary or middle school students. Does that mean to forget about improvising until the student has a firm grasp of all of this music theory knowledge? NO! Students should begin improvising as early as possible, but the teacher must guide them to repertoire appropriate for their theoretical level. For example, it would be a terrible idea to expect a young student to improvise over John Coltrane's *Giant Steps* in which the chords change every 2–4 beats and the key centers change just about as fast. Instead, the teacher must select a more appropriate tune such as Miles Davis's *So What*. As I discuss in Chapter 2, this tune uses only two different scales, and the chords change no faster than every 8 bars. Elementary aged students can learn the two scales and practice various patterns over those two scales and begin improving over *So What* fairly quickly.

. . . something that everyone can do.

In the past, people used to think that either you had it or you didn't. Today, it is obvious that improvisational skill isn't something you are born with. Through listening and ear training, music theory knowledge, and technical development, anyone with a willingness to improvise can be successful. I shouldn't understate the word "willingness" in the previous sentence. It takes a tremendous amount of hard work to become a great improviser, and improvisation is a skill that slowly improves over time. Developing improvisational skill is a marathon and not a sprint! As I discuss in this book, technology can play a huge role in making the journey easier and more enjoyable.

. . . for every genre of music.

From classical to jazz, country to rock, and pop to hip-hop, improvisation can be integrated into every genre of music. At the heart of all jazz music, improvised sections often account for 75% or more of a given recording. Jazz artists generally state the melody at the beginning and end of a performance and take turns improvising throughout the entire middle section. Modern hip-hop artists often improvise, calling it free-styling. Rock and country songs, while more focused on melodic vocal sections such as the chorus, verse, and bridge, will often include short improvised solo sections. In the classical world, young classical musicians are notorious for shying away from improvising. So much of our modern classical musician training focuses only on playing the notes on the page; however, this was not the case in the past:

> Performers in the Baroque era were always expected to add to what the composer had written. For example, keyboard players realized figured basses by improvising chords, arpeggios, and even counterpoints. Vocal and instrumental solo performers applied skill, taste, and experience to realize the full effect of the music by means of ornaments and embellishments.[2]

It is interesting that Bach, Mozart, and Beethoven were all outstanding improvisers. Bach was known as the greatest organ improviser of his day. He would even write out multiple versions of his inventions to illustrate how a student could improvise over the structure of the piece. Mozart was first recognized as an improviser in Vienna, and it wasn't until approximately 10 years later that he became known for his compositions. In an attempt to impress Mozart, a young Beethoven improvised on themes provided by Mozart at a party. All three of these "Mount Rushmore" composers placed improvisation at the heart of their musical training and performance. Today, most classical musicians perform cadenzas and preludes as written in the manuscript, but this was not the case in period performances. These sections were generally improvised by the performer. The great saxophonist, Branford Marsalis, did just this in his 2001 album, *Creation*. Marsalis is best known as an improvising jazz saxophonist and he carried his improvisation skills into the cadenzas of standard classical saxophone works such as the Ibert *Concertino da Camera*. As a student, I studied and performed the Ibert and it was brilliantly refreshing to hear an improvised cadenza.

For more interesting notes about classical musicians and improvisation, visit https://ericbarnhill.wordpress.com/facts-about-improvisation/.

Developing Improvisation in Musicians

Every master carpenter has a tool box, and he or she has spent years learning how to use each tool effectively. The carpenter has to use a variety of tools to create each new project. As a child, I used to love watching Norm Abram on *The New Yankee Workshop* on PBS create incredible cabinets, tables, chairs, and more out of wood. Norm had more tools in his workshop than anyone I could imagine, at least more than my dad! He always knew which tool would be most

2 Donald Jay Grout and Claude V. Palisca, *A History of Western Music* (New York: Norton, 2001), 361.

appropriate for each step of the project, and he effortlessly went from router to table saw to jigsaw.

Every improviser needs to pull from his or her musical tool box when improvising. The more tools, the greater is the variety of sounds available to the improviser. But remember, you can build a lot with basic tools such as a screwdriver, hammer, saw, and drill! The most essential tools to develop are scales, chord tones, and melodic motifs (called *licks* in the jazz world).

1. **Scales:** Although many scales exist, my top five most important scales are major, mixolydian, dorian, harmonic minor, and blues. Major scales are the foundation of almost all music and it is essential to learn all 12 major scales. Mixolydian (major scale with a flat 7th) and dorian (major scale with a flat 3rd and flat 7th) are derived from the major scale. For songs in a minor tonality, the harmonic minor scale (major scale with a flat 3rd and flat 6th) is essential. Finally, the blues scale is the swiss army knife of scales that can function in major or minor tonality and get you out of any tough improvisation situation.
2. **Chord tones:** The improviser will typically work with a lead sheet that includes the melody of the tune plus the accompanying chords. The top five most important chords to learn are major 7th (Maj7), minor 7th (min7), dominant 7th (dom7), minor with a major 7th (min M7), and minor 7th flat 5 (min7b5 or ø7). The player should be familiar with the 1-3-5-7 of each chord. Practice playing the chords ascending and descending on each chord tone.
3. **Melodic motifs:** Melodic motifs are the *lingo* of the musical world. Just as a writer takes common words and phrases and combines them in ever new ways, musicians borrow, develop, and manipulate melodic motifs to create freshly improvised solos. I suggest starting by borrowing your favorite phrases of great improvisers. For example, my two favorite saxophonists are Dexter Gordon and Sonny Stitt and, while listening to a recording, I will often pause the recording to write out one of their tasty licks. Another way to develop *lingo* is to write out cool phrases you discover through improvising. Sometimes, interesting phrases just come out and it a good idea to capture the phrase before you forget it. Who knows, maybe that phrase ends up becoming the basis of your next musical composition! So through a combination of borrowing phrases from great improvisers and developing your own, the improviser develops an aural catalog of cool licks that can be used throughout the improvisation process.

TIP

Purchase a spiral-bound manuscript book and notate melodic motifs that you borrow or create. You will soon develop a large library of improvisational *lingo* in this book.

The best way to learn how to effectively use all of your tools mentioned above is to JUST DO IT! The more you improvise, the more comfortable you will become and the better you will be at improvising. In Chapter 2, Auto Accompaniment Software, and Chapter 4, Music

Production Software, I detail how to generate custom accompaniments so you will have a backing track to improvise along with. Improvising with a playalong track will help train your ear to identify sounds you like and those that don't really fit. Always practice at a slow tempo so your ear, brain, and fingers can keep up. In Chapter 3, Notation Software, I show you how to use computer notation software to generate technical exercises and write out your own solo (sort of like pre-planned improvisation). This process allows the musician to work out ideas on paper/screen before trying to execute them in the moment. Practice these exercises and solo ideas in the abstract to allow them to flow out of you in the moment during your actual improvisation.

When you are practicing improvisation, keep the following in mind:

1. **Keep your place in the song by focusing on the form of the piece.** Is the song a 12-bar blues, 32-bar AABA, or something altogether different? Knowing the form can help keep you from getting lost while improvising or get you back on track if you do get lost.
2. **Play the "right" notes.** Throughout this book, I'll continually mix in theory and improvisation strategies to limit your tonal palette. Very often it is best to limit yourself to a particular scale or set of chord tones as a beginning improviser. Continually self-monitor while you practice to be sure you stay "in the box."
3. **Use repetition and sequencing.** Obviously repetition means to play phrases more than once. As humans, we crave repetition in music. Select any melody and I'm sure you'll find many instances of repetition. If it works in your favorite melody, it will work in your improvisation. Consciously focus on repeating musical phrases to keep some logical sense to your on-the-spot composition. When you repeat a melodic idea on a different note or in a different tonality this is called sequencing. In the song "I'll Remember April" the first four bars start in G major and the second four bars move to G minor. It is especially effective to perform a given phrase in major and then sequence it to minor in this situation. In "Satin Doll" the first 2 bars use a ii-V progression in the tonic key while bars 3–4 use a ii-V progression a whole step up. Again, it is very effective to play a given phrase in the tonic key and then again a whole step up. In both situations you satisfy the human need for repetition while also adjusting your improvised melodic idea to fit the harmony of the tune.
4. **Practice at slow tempos.** There is simply no need to increase the tempo past your comfort zone. Improvisation is difficult and it takes time to coordinate your ear, brain, and fingers. I will continue to recommend starting slow throughout this chapter and this entire text. All of my private lesson instructors from elementary through college have reminded me that "Perfect practice makes perfect." Don't bother practicing mistakes or pushing the tempo past the limits of your ear, brain, or fingers. Make the most of your practice time by getting it right at a comfortable tempo in the practice room so you will ultimately have more success on stage.

Finally, you must possess *self-confidence.* Whether you call it mojo or swagger, self-confidence is one of the keys to improvisation. Your playing must be rhythmically confident to fit in the groove or to create exciting syncopation. Your playing must be melodically confident to fit in the harmony. Wimpy playing never leads to a quality performance. Bill Evans, the great jazz pianist, once said, "There are no wrong notes, only wrong resolutions." Hear the music in your head, execute it on your instrument, and, if you miss, you are probably only a half step away from a nice resolution.

Check out this link for more interesting "no wrong notes" quotes from jazz greats:

http://www.bryanjudkins.com/post/149429495/there-are-no-wrong-notes-in-jazz-only-notes-in.

Repertoire Guidelines

Selecting appropriate repertoire for the improviser's level is of utmost importance. As I mentioned, the three limitations of any improviser center around technical ability, ear development, and music theory knowledge. The four types of tunes best for beginning to intermediate improvisers are vamp, modal, blues, and ii-V-I tunes.

Vamp

A vamp is a repeated musical figure (such as a guitar riff). This vamp can serve as the basis for the entire song or just a section of the song. Sometimes only the improvisation section of the song is based on a vamp. Stevie Wonder's "Superstition" is a great example of a vamp tune. This tune starts with a basic drum groove followed by the iconic 2-bar clavinet vamp line. The rest of the tunes builds on the vamp adding vocals and horns. The song eventually moves to a turnaround section that leads directly back the vamp. Over a tune such as "Superstition," the improviser doesn't need to worry about getting lost in the chord changes and he or she has a very limited amount of harmony to deal with.

Modal

Instead of using a complex chord progression, modal tunes use chords based on musical modes generally from the major scale. These tunes will often feature very few chords and the chords will often change relatively slowly. Miles Davis's album *Kind of Blue* features many modal tunes with very simplistic harmony. This allows the beginning improviser to focus on a small number of scales with very clear changes in harmony throughout the tune.

Blues

The blues is a classic form common in jazz, rock, and obviously blues songs. These songs feature a form most commonly 12 bars in length with as few as three chords. While there are many variations on the blues progression, the core of just about all blues songs is the I, IV, and V chords. There are countless examples of blues songs including James Brown's "I Feel Good," Pink Floyd's "Money," and Duke Ellington's "C Jam Blues." The blues allows the musician to clearly begin using various tools to create a more interesting improvisational experience. At the most basic, an improviser can use only the blues scale of the tonic key throughout the entire song. Taking a step forward, the improviser can use the mixolydian mode over each chord. So in the key of C, the improviser would perform C mixolydian over C7, F mixolydian over F7, and G mixolydian over G7. Finally, the improviser can explore the chord tones of the I, IV, and V chords. By combining all three tools (blues scale, mixolydian mode, and chord tones), the improviser really starts down the road of creating more musically interesting improvisational material. I further explain the basic blues progression and the appropriate improvisation tools in Chapter 2, Auto Accompaniment Software.

ii-V-I

As students gain comfort improvising over vamp, modal, and blues tunes, they are ready for ii-V-I based songs. The ii-V-I progression serves as the foundation of most jazz standards and many pop songs. Although ii-V-I songs often include relatively quick-moving harmony compared to vamp, modal, and blues songs, it is possible to bracket chunks of the tune into a single scale making it easier for a young improviser to solo over this sort of tune. For example, in a 4-bar section that includes Dmin7, G7, and CMaj7, the improviser can bracket all three of these chords under the C major scale umbrella. Tunes with the ii-V-I structure also allow the musician to push further into chord tone arpeggiation. Often these tunes will include sections in a minor tonality, forcing the improviser to explore the harmonic minor parent scale. I further explain the mechanics of improvising over a ii-V-I based song in Chapter 2, Auto Accompaniment Software.

Classroom Technology

When you are teaching in a classroom setting it is essential for your students to be able to see and hear your computer/iPad. This allows the students to follow along at their seats. Some of the gear discussed in this section allows the teacher to easily communicate and share audio with the entire class or individual students for one-on-one instruction.

Seeing the Teacher Device

When I began teaching public school in 2002, it was rare for regular classrooms (let alone music classrooms) to be outfitted with a large computer display device such as an LCD projector with screen or interactive whiteboard. Today, this technology is finding its way into more and more classrooms. At Haverford High School, we have LCD projectors in just about every classroom. At Manoa Elementary, my son's elementary school (and my alma mater!), Smart Boards were placed in every classroom when the school was rebuilt in the early 2010s.

LCD Projectors simply project the teacher's computer/iPad to a screen for large group instruction. If the teacher's computer is in close proximity to the screen, an LCD projector works just fine. In my classroom, my computer is five feet from the screen so I can easily click on the computer and then walk to the projector screen if needed. If the computer is farther from the screen or in the back of the classroom, a wireless keyboard and mouse can allow the teacher to provide instruction from the front of the room near the projector screen. Be sure to create a comfortable teaching position with you at the front of the room facing the students.

The most common interactive whiteboards that I see in schools are SMART Boards and Promethean Boards. Both essentially accomplish the same task. The teacher's computer/iPad is connected to the built-in LCD projector on the interactive whiteboard. After running a quick configuration and setup in the interactive whiteboard software on the teacher computer/iPad, the teacher can now work with his or her device from the actual whiteboard in the front of the classroom, eliminating the need for direct access to the computer mouse/keyboard. I find interactive whiteboards especially appropriate for elementary-aged students although they can be used successfully with students of any age.

When connecting your device you will need to purchase the necessary video cable to connect to your LCD projector or interactive whiteboard. Most commonly you will need a VGA cable. Some newer displays require an HDMI cable similar to the cable used in most cable television boxes. Also check the video output of your computer or device. Since many Windows computers have VGA ports, you will likely only need a single VGA cable with no adapter. Macintosh computers and iPads almost always require an adapter to connect to the VGA cable. For iPads you will need a "lightning to VGA" (iPad 4th generation and newer) or "30-pin dock connector to VGA" (iPad 3rd generation and older) adapter. For Macintosh computers it will all depend on your computer model. Current Macs require a "mini display to VGA" adapter while older Macs used a number of different video output ports. These adapters typically cost $30 to $50.

Follow this link for more information on iPad video adapters:
https://support.apple.com/en-us/HT202044.

Follow this link for more information on Mac video adapters:
https://support.apple.com/en-us/HT201853.

Hearing the Teacher Device

Quality audio playback is needed in any music classroom situation. In the simplest setup, students will connect stereo headphones into their devices and the teacher will have speakers to play back sound from his or her device to the class.

When selecting a set of speakers for teacher device playback you will have a choice between active or passive monitor speakers. Active monitors are self-powered speakers that have a power cable that plugs into a wall outlet. You will need a power supply in close proximity to the active monitors. Passive monitors do not require a power supply; however, you will need to purchase a power amplifier to power the speakers. I suggest purchasing a Crown power amplifier; these amps have been around for years and are the gold standard of power amplifiers. You must match the power amplifier to your passive speakers, so check the power (watts) and resistance (ohms) of your speakers and power amplifier when purchasing so you can match the two. The general rule of thumb is to purchase a power amplifier at least 50% more powerful than your speakers at the given resistance. For example, if your speakers are rated at 600W continuous power at 8 ohms, you'll want a power amp rated at approximately 900W at 8 ohms. Refer to this YouTube video for more detailed information on amp and speaker matching: https://www.youtube.com/watch?v=pUou_noD1Gc.

TIP

Matching amplifiers and speakers can be confusing. I suggest enlisting the assistance of a qualified sale representative from a reputable dealer before making this important purchase.

For audio playback, my preference is to use a lab management system such as the Korg GEC to make sharing audio more seamless and flexible. The newest model is the Korg GEC5. See Figure 1.1 for an image of the Korg GEC5 teacher unit and Figure 1.2 for an image of the Korg GEC5 student communication interface. To set up the GEC5, you must install a student communication interface (SCI) at each student computer station. Sound output from the student computer, keyboard, and headset microphone is routed to the SCI. A CAT-5 ethernet cable links the SCI to the GEC5 teacher unit (GEC5-TU) at the front of the room near the teacher computer. Similar to the SCI setup, sound from the teacher keyboard, computer, and headset are routed to the GEC5-TU. The GEC5-TU can also output sound to room speakers and the teacher can even record the audio output of the GEC5-TU to capture student performances. The main reason to use the Korg GEC5 is that the teacher has supreme flexibility in audio playback, and it is a piece of gear that actually allows a classroom to *function*. Over my 13 years of teaching music technology both in high school and higher education, the one indispensable piece of gear that makes my classroom *work* is the Korg GEC. Administrators are always amazed at how quiet my classroom is and how efficiently I can address student needs and differentiate my instruction. Students can work independently, interact one on one with the teacher, share audio with the class, and work collaboratively with a peer. The GEC

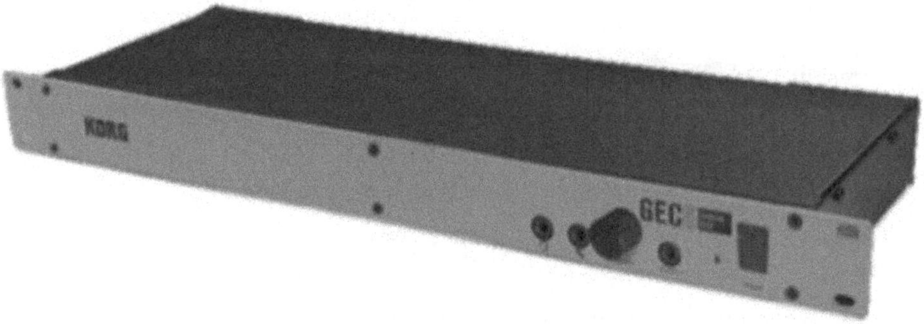

Figure 1.1 Korg GEC5 teacher interface.

Figure 1.2 Korg GEC5 student communication interface.

is the audio nerve center of my classroom and provides the perfect teaching environment for all audio-based activities. The teacher can manage all of the sound right from the front of the room or, with the addition of a wireless router, the teacher can manage the classroom while walking around the room with an iPad or Android tablet.

Digital Instruments

In a music classroom, acoustic instruments can be troublesome simply because they actually produce audible sound. This can make teaching improvisation in a classroom setting very difficult unless there are separate practice rooms or spaces for students to play their acoustic instrument with a computer/iPad that won't disturb classmates. Many teachers opt for digital instruments to solve this issue.

The piano keyboard is the most common solution. When purchasing a piano keyboard you will have a choice between a model with built-in sounds and one without built-in sounds (called a controller). Controllers are generally much cheaper (approx $50–$200) but they must be connected to a computer with software instrument sounds (such as a Mac with GarageBand installed) to produce any sound. Keyboards with built-in sounds can cost significantly more than controllers but they produce sound without the need to connect to a computer/iPad. My preference is to always purchase keyboards with some built-in sounds if your budget permits.

Although keyboards are the most common digital instruments, they are by no means the only option. The EWI (see Figure 1.3) is an electronic wind instrument meant to mimic a saxophone, clarinet, or flute. Although I'm an experienced saxophonist, I found that working on the EWI takes a good bit of practice since it doesn't react exactly like an acoustic wind instrument. With practice, the EWI can be a good option for woodwind players. Electronic drum sets (see Figure 1.4) made by companies such as Roland and Alesis make use of various pads that send MIDI signals to a computer software instrument or an included sound module. Still other digital instruments such as the Novation Launchpad (see Figure 1.5) look much less traditional and can allow students to improvise without prior knowledge of a traditional instrument.

When connecting a digital instrument your computer you may need a MIDI interface. A MIDI interface allows your digital instrument to communicate to your computer/iPad through a USB cable. The M-Audio UNO is my MIDI interface of choice because of its small size, ease of use, and relatively low cost (approx $39). See Figure 1.6 for an image of the

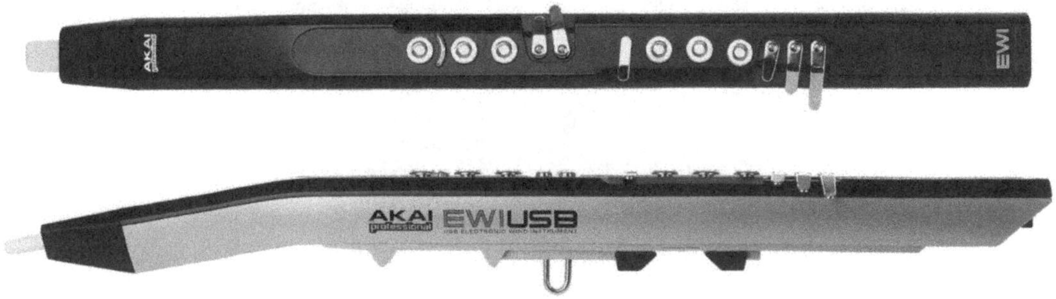

Figure 1.3
Akai EWI MIDI controller.

Figure 1.4
Roland V-Drums MIDI controller.

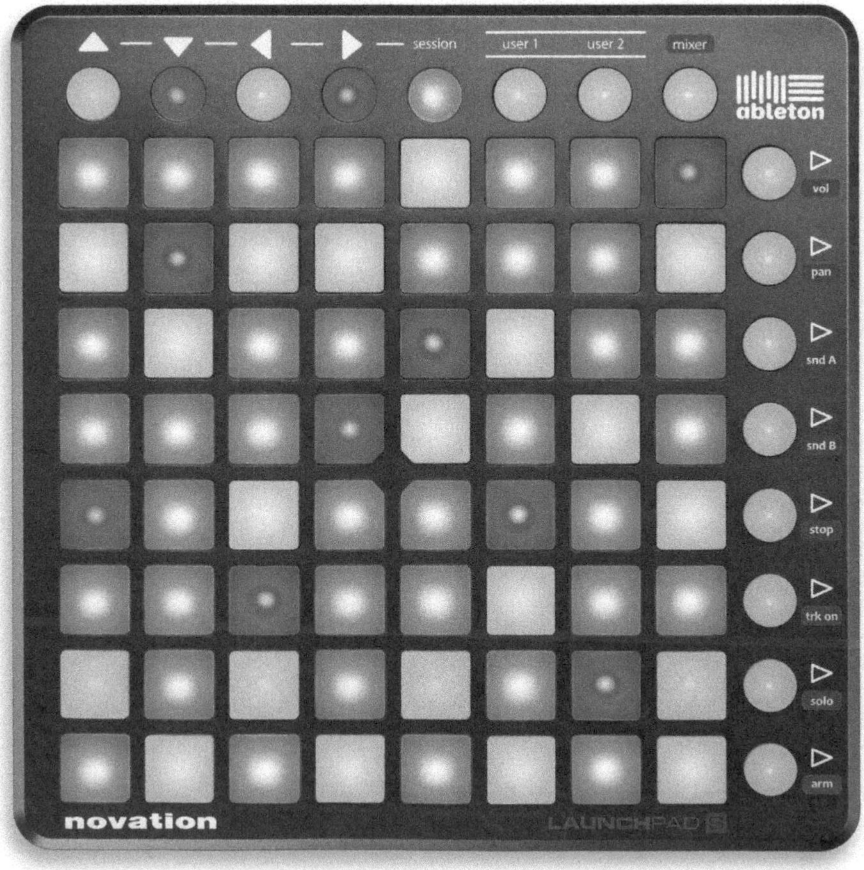

Figure 1.5
Novation Launchpad MIDI controller.

Figure 1.6
M-Audio UNO MIDI interface.

M-Audio UNO. Connect the USB end of the UNO into your computer/iPad and the MIDI ends to your digital instrument's MIDI in and out. If your digital instrument has a direct USB cable out of it, your digital instrument most likely has a built-in MIDI interface and you will not require a separate MIDI interface. If you are using an iPad you will also need a separate adapter called the "Lightning to USB Camera Adapter" (for 4th generation iPad and newer) or "Apple iPad Camera Connection Kit" (for 3rd generation iPad or older). I know it is odd to call this adapter a "Camera Kit" but it really just converts your iPad's input port to a USB port.

How Can Technology Facilitate the Development of Musicians' Skill in Improvising?

This text covers auto accompaniment software, notation software, and music production software as well as web resources for listening, posting, and organizing material for students. Below is a summarized list of various activities that will be expanded on throughout the remainder of this text:

- Create play-along recordings so students can hear their improvised ideas with the rhythm and harmony.
- Use software to generate a solo, create appropriate notation, and print for practice.
- Notate scale and chord tone exercises.
- Notate melodic motifs and transcribe improvised solos.
- Transpose notated exercises for various instruments and ranges.
- Have students write out a solo ahead of time and then perform it on their instrument.
- Edit audio recordings to highlight important aspects of the performance.
- Develop listening playlists on YouTube and/or Spotify.
- Develop a series of podcast episodes that illustrate beginning improvisation concepts.
- Develop a website to organize all of your materials and so students can easily access files from home.

Existing Materials and Methods

My first use of technology with improvisation came with the Jamey Aebersold play-along books (see Figure 1.7). Each book includes a handful of jazz tunes transposed for concert, bass clef, E♭, and B♭ instruments plus a CD accompaniment. The accompaniment usually includes drums, bass, and piano so a melody instrument such as saxophone can perform the melody and improvise over the tune. The accompaniments are typically mixed with the bass in the left channel only, piano in the right channel only, and drums in both channels. Because of this setup, the improviser could mute the piano by unplugging the right speaker or the bass by unplugging the left speaker. While many volumes of Aebersold books focus on tunes, others focus on common chord progressions or scales that improvisers often encounter, and Aebersold creates exercises out of each in all 12 keys. Jamey Aebersold has published more than 100 improvisation play-along books and various other practice aids. Explore the full collection of Aebersold materials at http://www.jazzbooks.com/.

Figure 1.7 Jamey Aebersold's Volume 54: *Maiden Voyage* play-along book.

Listed here are some of my favorite Aebersold volumes:

- Volume 3: The ii/V7/I Progression
- Volume 7: Miles Davis
- Volume 25: All-Time Standards
- Volume 31: Jazz Bossa Novas
- Volume 37: Sammy Nestico
- Volume 54: Maiden Voyage
- Volume 50: The Magic of Miles Davis
- Volume 65: Four and More

In terms of band methods, the two primary resources are *Essential Elements Jazz Method* published by Hal Leonard (see Figure 1.8) and *Standard of Excellence Jazz Method* published by Kjos (see Figure 1.9). Both methods include improvisation studies and short method book-style exercises;

Figure 1.8
Essential Elements Jazz Method.

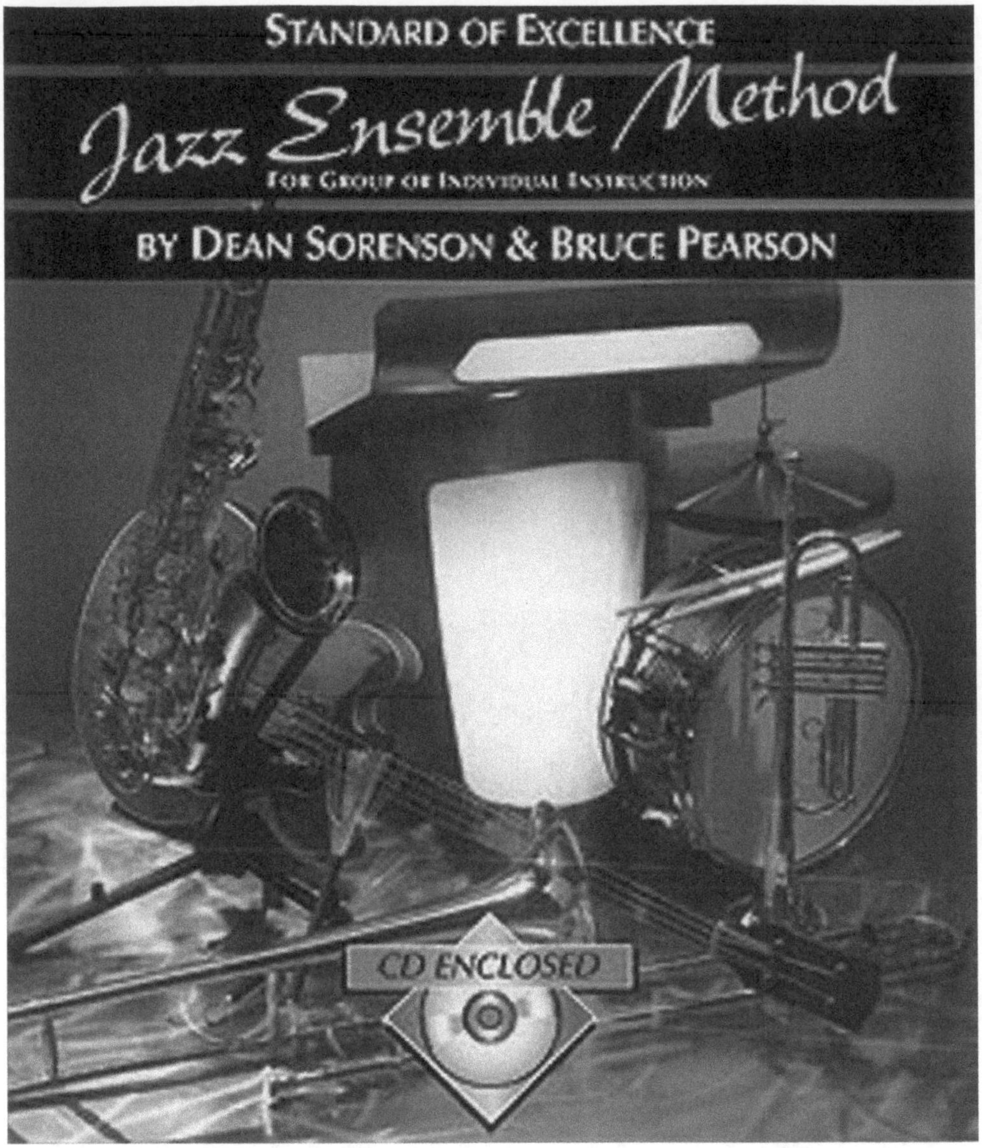

Figure 1.9 *Standard of Excellence* Jazz Method.

and full band arrangements and recordings of each exercise/arrangement are available on the included CD. The Hal Leonard series includes the ensemble method, *Jazz Standard* and *The Blues* play-along books, and a handful of grade 1-2 ensemble works appropriate for small or large ensemble. The play-along materials include computer software that allows the student to adjust the tempo of the tracks. The Kjos series contains the basic jazz ensemble method, advanced jazz ensemble method, three grade 1 full band arrangements, and two grade 2 full band arrangements.

- For more information on the *Essential Elements Jazz Method* published by Hal Leonard visit http://www.halleonard.com/ee/band/jazz.jsp.
- For more information on the *Standard of Excellence Jazz Method* published by Kjos visit http://www.kjos.com/sub_section.php?division=6&series=80.

SmartMusic is practice software developed by MakeMusic (the creators of *Finale* music notation software). See Figure 1.10 for an image of SmartMusic software on a Mac/PC. SmartMusic

Figure 1.10

SmartMusic software for Mac/PC.

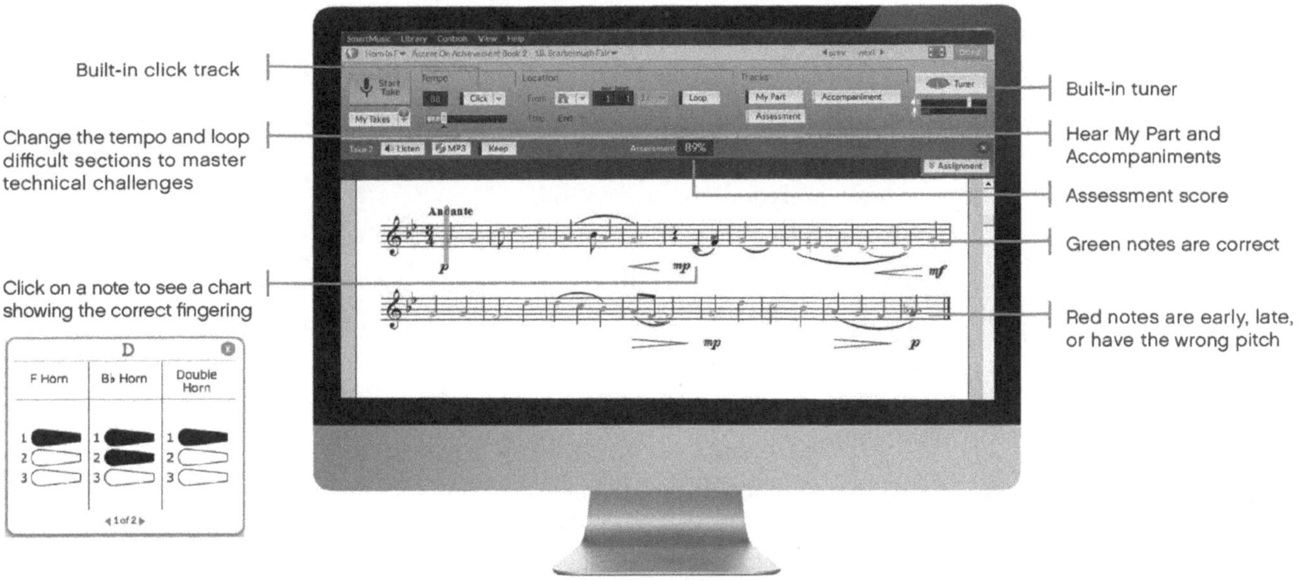

is available as a yearly subscription, and students can purchase it for only $40/year. Educator subscriptions cost $140/year and include access to a classroom interface so the teacher can easily keep track of student progress. Both subscriptions include identical repertoire material and software. Although $40 may seem like a significant yearly investment for some students, the SmartMusic library is extremely extensive and includes accompaniments for scale/arpeggio exercises, ear training, sight-reading, standard solo repertoire, full band/orchestra/choir repertoire, and every standard method book with music on-screen. Many students skip purchasing a paper method book and opt to subscribe to SmartMusic for about $25 more than a paper method book since SmartMusic already contains the full method book plus hundreds of additional resources. Here's some additional information from http://www.smartmusic.com/:

> Student practice is focused because they receive immediate feedback while listening to their performance and seeing the correct and incorrect rhythms and pitches on-screen. It also allows the teacher to provide students with the individual instruction and customized feedback needed to get better faster. Students are also able to hear their part in context with SmartMusic's professional background accompaniment, giving them a pitch and rhythmic reference when practicing at home.

In addition to the included repertoire, Finale software owners can create custom materials by exporting any Finale file as a SmartMusic file and then importing that new file into SmartMusic. When exporting as SmartMusic, you will have a choice of labeling the file as *Assessement, Solo,* or *Ensemble.* An *Assessment* file grades the student performance. A *Solo* file follows student tempo changes and permits transposition. An *Ensemble* file allows the student to mute selected instruments/voices.

After you purchase a SmartMusic subscription, download the SmartMusic application from http://www.smartmusic.com/support/downloads/. Login using your username and password and select *Find Music* on the left side of the screen. For practice focused on improvisation, I suggest exploring the following categories:

1. Exercises:
   a. **Scales:** This section includes major, minor (natural, harmonic, and melodic), chromatic, whole tone, pentatonic, and diminished scales in a variety of patterns. Major and harmonic minor scales will serve as the basis for much of our vamp and modal improvisation.
   b. **Arpeggios:** Knowing chord tones will eventually be important for students to take the step beyond simply improvising on a given scale. This section includes exercises for all chord qualities; however, I would focus primarily on the major, minor, and dominant 7th chords.
   c. **Jazz:** This section includes exercises on the blues scale, ii-V-I scales/arpeggios, and iii-VI-ii-V-I scales/arpeggios. The blues progression and ii-V-I tunes will be used throughout this text.
   d. **Blues Licks:** This section is all about ear training! SmartMusic plays a phrase and you respond with the same phrase—and remember: SmartMusic will evaluate your performance. Explore the keys of F, B♭, C, G, and E♭ (the most common keys for blues progressions) and the variety of feels/styles (slow, medium, fast, swing eighths, straight eights, New Orleans, and rock).
   e. **Play by Ear:** Although this section is labeled "classical ear training," any improviser can benefit from exploring the seven levels offered here.
2. **Jazz Improvisation:** This section includes play-alongs with music on-screen from the Alfred MasterTracks series, Gordon Goodwin's Play-along series, SmartMusic Improv series, and Wynton Marsalis series. Each tune includes the chord progression, and many include the melody on-screen. Students can transpose the tunes to any key and also display patterns such as 1-3-5-7 or 5-3-2-1 for each chord as a practice aid. When using the patterns, students can also select "Learn Chords" to omit 1-3 notes from the chord pattern. Perhaps the coolest parts of these materials are the piano, bass, and drum transcriptions. These parts are available from the drop-down menu in the upper left corner and display the exact transcription of each accompaniment part. This is a great way to actually show a chordal instrument player how to voice a chord, a bass player how to walk over a progression, or a drummer how to play a particular groove or fill.

iPad Connection: SmartMusic for iOS

Available since 2014, the long awaited SmartMusic app brought the power of the desktop application to the iPad tablet (see Figure 1.11). Although the iPad version lacks some features

Figure 1.11
SmartMusic application for iPad.

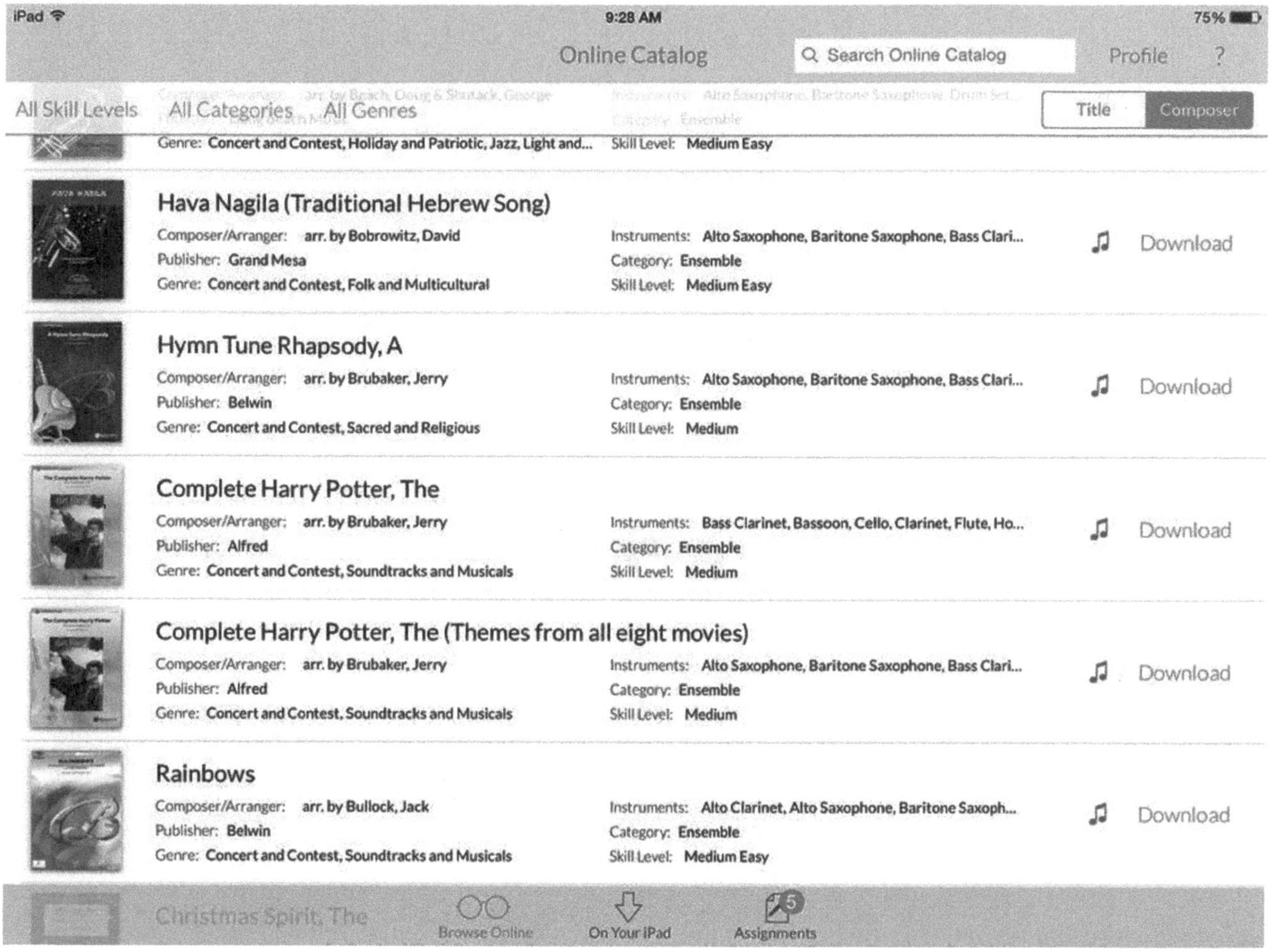

and does not include 100% of the library material from the desktop version, having the most-used SmartMusic material on my iPad is incredibly convenient.

1. Download and install the SmartMusic app from the Apple App Store on your iPad or via iTunes on your desktop computer.
   a. Download link: https://itunes.apple.com/us/app/smartmusic/id638851328?_ga=1.26756789.940862473.1442405231.
   b. For more information and video demonstrations visit http://www.smartmusic.com/mobile/.
   c. For the online SmartMusic app user manual visit http://www.smartmusic.com/onlinehelp/mobile/.
2. Tap *Profile* in the upper right corner. Select your primary instrument and any additional instruments you'd like available in all musical activities in the app. You can only select a primary and four secondary instruments.

TIP

When you select an instrument you will also have the other instruments in that family available. For example, if I select *Alto Saxophone* in my profile, I will automatically have soprano, alto, tenor, and baritone saxophones available in the musical activities.

3. Tap *Find* at the bottom of the screen.
4. Tap *All Categories* and select *Exercises.*
5. Swipe up until you locate *Jazz* and tap it.
6. In the next few screens tap *Blues* > *Quarter Notes* > *1 Octave Up*.
7. You are now at the main playback screen (see Figure 1.12). Set your instrument in the upper left corner.

Figure 1.12
SmartMusic app for iPad main playback screen.

8. Tap the *Metronome/Tuner* icon in the top center of the screen. Adjust the tempo to a comfortable speed and tune your instrument.

TIP

Tap *My Part* to hear the notes on-screen along with the accompaniment.

9. Tap the *Microphone* icon to begin playback. As you play, SmartMusic is listening to and evaluating your performance. When you have finished the take, you will notice red and green notes on the staff. Red notes indicate wrong pitches/rhythms while green notes indicate correct ones. SmartMusic will also give you a percentage for each take.

TIP

Tap the screen during playback to pause or stop the take.

10. Tap the *Microphone* icon again to do additional takes.
11. Tap the *Headphone* icon to listen to your performance. As the file plays, you can adjust the balance between your microphone and the accompaniment.
12. Select the best take, tap the *Share* icon in the upper right corner, and tap *Mail* to email a recording of your performance.
13. Tap the letter of the key and select a new key. Continue practicing this scale in all 12 keys.

TIP

Tap the *Cycle* icon to the right of the *Microphone* icon in the top center to cycle the exercise to all 12 keys. See Figure 1.13.

Figure 1.13
Cycle options in the SmartMusic app.

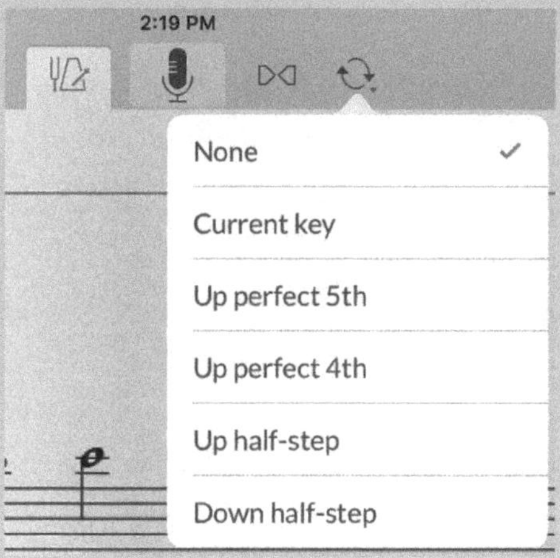

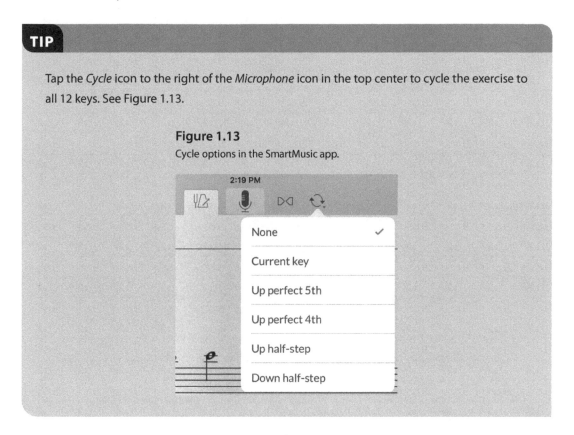

14. Tap *Back* in the upper left and continue to tap that corner to back out of the *Blues* exercises.
15. Explore the other exercises available under the *Jazz* category including *Blues*, *II-V-I Scales*, *II-V-I Arpeggios*, *III-VI-II-V-I Scales*, and *III-VI-II-V-I Arpeggios*. As previously mentioned in this chapter, these scales and chord progressions are extremely common and very important to your development as an improviser.
16. Navigate to the main *Exercises* page of the app and select *Blues Licks* > *Blues in B♭* > *Rock Blues in B♭*.
17. Tap the *Microphone* icon to begin playback. In this ear training activity the goal is to repeat the computer generated part exactly. As usual, SmartMusic will evaluate your performance and the same listening, sharing, cycling, and key options from the previous exercise are available.
18. Tap *Back* in the upper left and continue to tap that corner to back out of the *Blues in B♭* activity. Explore the other blues ear training available under *Blues Licks* including *Blues in F*, *Blues in C*, *Blues in G*, and *Blues in E♭*.

TIP

The other exercise categories most applicable to improvisation include *Arpeggios, Intervals, Play by Ear,* and *Scales*. I suggest incorporating one of these exercises played in all 12 keys in your daily practicing. Remember that technical ability is a limiting factor in improvisation and it is important to constantly foster your technical development.

19. Back out to the main *Find* area of the app and tap *Reset* in the top row to bring everything back to default value.
20. Under *All Categories* select *Jazz Improvisation.*
21. Locate *Alfred MasterTracks, Blues* and open *T's Blues*. This song follows a 12-bar blues form with relatively basic chords. I delve deeper into the blues progression in Chapter 2 of this text.
22. Practice playing the melody using the music on-screen. There is ample space to improvise throughout the middle of the tune.

TIP

Unlike the desktop version, the iPad version of SmartMusic currently does not display chord patterns.

23. Tap the instrument selection area in the upper left corner and swipe up to reveal the piano, bass, and drum transcriptions. Invite rhythm section players to explore these transcriptions to help their comping technique.
24. Explore the other repertoire available under *Find* > *Jazz Improvisation* including *Alfred MasterTracks* (*Blues* and *Fusion*), *Rockin' Blues, Straight-Ahead Blues,* and *Strayhorn & More*. I particularly enjoy the *Strayhorn & More* collection that includes some very well-known tunes performed by the Duke Ellington Orchestra with wonderful melodies and chord progressions. Improvising on these tunes will be a challenge for intermediate to advanced improvisers and are not suitable for very beginning improvisers.

Breaking the Ice: Percussion-Only Improv Jam

[▶] *See Example 1.2 on the companion website for an example audio recording of the "Percussion-only Improv Jam."*

Imagine going to a party with people you never met before. Things generally feel awkward until one person steps up and introduces himself or herself, striking up a conversation. It then takes time to feel comfortable with that new person before you start revealing details of your personal life. When improvising, students need a chance to "break the ice" in a non-threatening environment with a supportive teacher and peers. The *Percussion-only Improv Jam* can do just that in a piano lab environment with the Korg GEC5.

1. Using the Korg GEC5 in a piano keyboard lab, group four students together.
   a. Select Group mode in the iPad or desktop GEC5 application.
   b. Click and hold on each student icon until it turns blue.
   c. After you have selected each student, click *New Group* in the lower left corner. See Figure 1.14 for an image of students grouped in quartets.

Figure 1.14
Students grouped in quartets in the Korg GEC5 application.

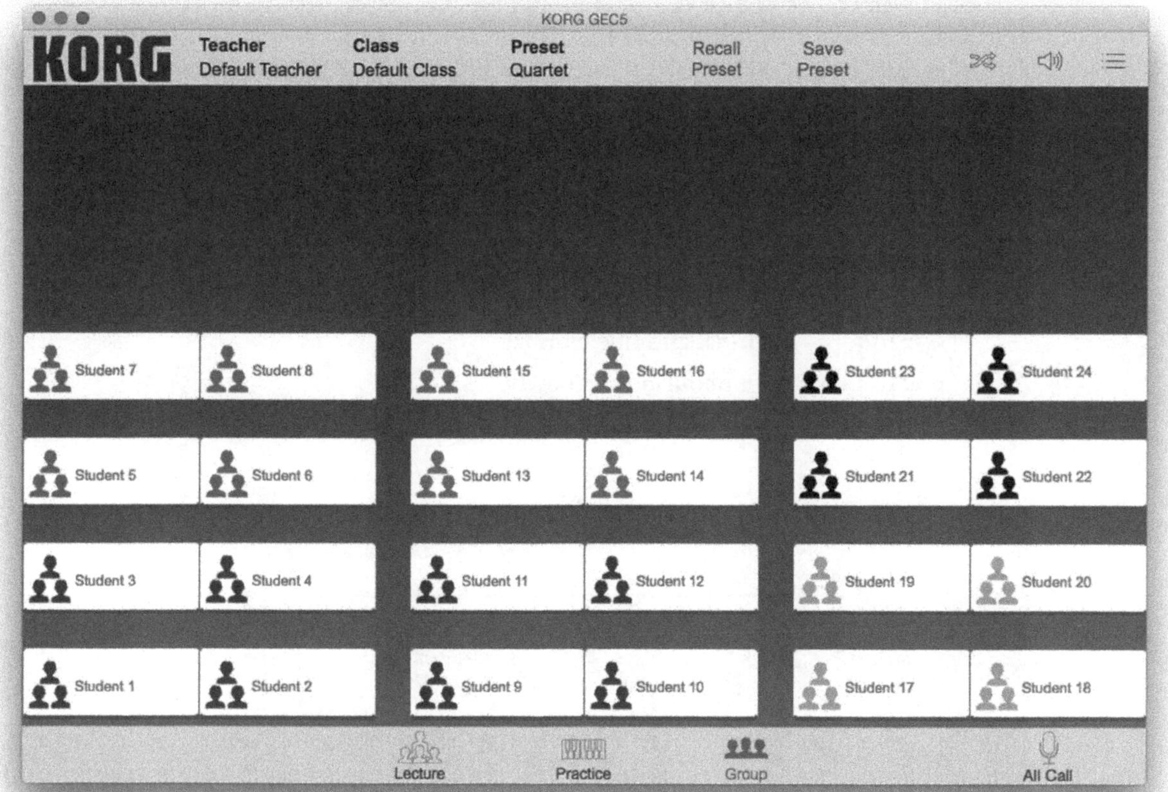

TIP

Instead of manually assigning students to groups, click *Recall Preset* in the top center of the screen and select one of the auto group options such as duet or quartet.

2. Direct students to open the General MIDI (GM) drum patch on their keyboards. This is typically found by going to the GM bank on the keyboard and going past

sound 128. Some keyboards require you to press the GM bank button twice to access the GM drums bank.

3. Explore the full set of drum/percussion sounds available in this patch. This GM drum bank is common to almost all keyboards and a full GM drum map is available here: http://www.mindwaremusic.com/SoundsetterHelp/scr/GMDrumMap.html .

TIP

I suggest downloading and printing a GM drum map for each student to reference throughout this project. Encourage students to do an Internet search for the percussion instrument they select for this activity to gain historical and contextual knowledge of the typical use of the instrument in acoustic musical ensembles.

4. Assign each student to a single key or group of keys. I suggest the following drum configuration to start. See Figure 1.15 for keyboard placement of these instrument sounds in a standard GM drum kit.
   a. Student #1: Kick drum and snare drum
   b. Student #2: Hi-hat and ride cymbals
   c. Student #3: Tambourine and cowbell
   d. Student #4: Congas

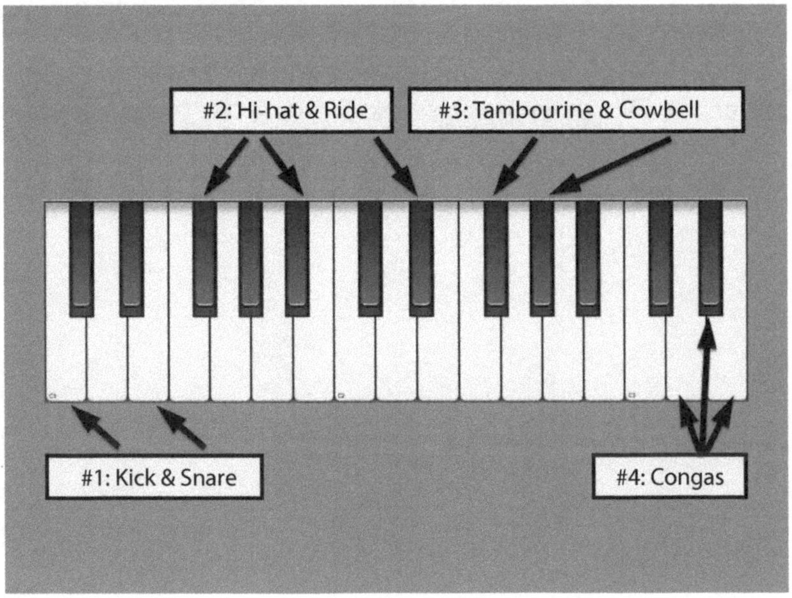

Figure 1.15 Suggested drum configuration for percussion-only improv jam activity.

5. Encourage one student to develop a 1–4 bar rhythmic groove. Layer in additional students one by one. Be careful to avoid overly busy rhythms so each instrument can have its own rhythmic place in the composition.

TIP

Students should notate their selected rhythm for future reference. If students are having difficulty transcribing their rhythm, they can perform the rhythm into GarageBand or Note Flight and view/print the computer generated notation. I delve into GarageBand music production software in Chapter 4 and Note Flight notation software in Chapter 3 of this text.

6. Once students have a basic idea of how each rhythm functions in the composition, encourage them to improvise some variations throughout the performance.

TIP

To create a rhythmic variation:

- Play the rhythm in half-time by doubling the length of each note.
- Play the rhythm in double-time by halving the length of every note.
- Drop out one or more of the notes.

7. For advanced groups, have each student select a different drum/percussion sound to develop a contrasting B section. Students can then perform the composition in an ABA form to add interest for a longer performance.

TIP

Although the Korg GEC5 is the most ideal way to execute this activity, consider modifying the task above to suit your specific classroom. Here are some suggested modifications:

1. **Keyboard lab without the Korg GEC:** Purchase a 4-input mixer (such as the Mackie 402VLZ4 - $99.99) and a headphone amplifier (such as Behringer MicroAMP HA400 - $24.99). Connect the L/mono output of each keyboard into the mixer inputs. Connect the main or headphone output of the mixer to the headphone amplifier input. Connect student headphones to each output of the headphone amplifier. Optional: You could record the performance using the tape output of the mixer connected to a handheld digital recorder (such as the Zoom H5 - $269.99) or computer with audio interface (such as the M-Audio M-Tracks II- $99.00).
2. **Classroom with miscellaneous percussion instruments:** Instead of using keyboard percussion, use acoustic percussion instruments such as shaker, tambourine, drums, cymbals, congas, etc. Consider recording improvised student performances using a handheld digital recording device or computer with audio interface.

Chapter 1 Review

1. Describe the importance of restrictions in improvisation.
2. Since improvisation is limited by technical ability, ear development, and music theory knowledge, identify strategies to begin teaching improvisation to young musicians. For example, select repertoire with limited chord changes such as Miles Davis's "So What."
3. Improvisation is present in every genre of music. Select a musician from any genre or period and describe the role of improvisation in his or her music.
4. Selecting appropriate repertoire is paramount. Identify at least two tunes from each category (vamp, modal, blues, and ii-V-I).
5. *Scenario:* Your school principal has funding available to enhance the technology currently in your classroom. *Challenge:* Create a budget for technology equipment you'd request to help teach improvisation in your classroom. I suggest including gear to see/hear the teacher's computer, share audio in the classroom, and any digital instruments or accessories you'd require.
6. Compare and contrast the four existing materials and methods discussed in this chapter (Aebersold Play-alongs, *Essential Elements Jazz Method, Standard of Excellence Jazz Ensemble Method,* and SmartMusic).

2 AUTO ACCOMPANIMENT SOFTWARE

Band-in-a-Box and iReal Pro

Band-in-a-Box is a very unique piece of software made by PG Music for Mac/Windows that actually creates a full accompaniment with piano, bass, and drums (and sometimes more) based on chords that can be entered in seconds. I first began using Band-in-a-Box in 1994 during my high school career. We had a single computer in a practice room (it was actually a storage closet for marching band uniforms, but we made it work as a practice room!) and the jazz students would go back there from time to time during concert band rehearsal to "shed" (practice) the "changes" (chord progression) of our current jazz ensemble tunes. Band-in-a-Box comes with a variety of styles so the user can create an accompaniment in virtually any style from classical to jazz, country to pop, rock to hip-hop, and more. In addition to using Band-in-a-Box to create improvisation backing tracks, I also use Band-in-a-Box to create custom accompaniments for general music classroom songs for use with preschool and elementary aged children. Doing this allows me to focus my attention on the musicality and vocal performance of my students while also giving me the option to add additional acoustic accompaniment (such as guitar or percussion).

Band-in-a-Box is a great example of a tool that doesn't get in the way of practice time because you can create an accompaniment so quickly. In this chapter, we use four different songs to delve into the core theoretical improvisation concepts needed to be successful as a new improviser. We use Band-in-a-Box as a tool to create accompaniments, vary styles, create a solo using only the QWERTY keyboard, and generate a solo using the soloist generator. In the "iPad Connection" portion of this chapter, I delve into iReal Pro, one of my favorite apps for iOS that is similar to Band-in-a-Box. iReal Pro offers practice features such as looping, transposing to new keys, transposing display for transposing instruments, style selection, and, my personal favorite, the iReal Pro Forum with thousands of tunes already created by users and ready for teacher/student use!

"E♭ Jam Blues"

Basic Blues Progression

The blues originated in southern African American communities in the 1800s. It is a uniquely American musical genre that combines African and European influences and has permeated just about every musical genre in America that has been created after it. We can clearly see the influence of blues melodies, form, and chord progressions in jazz, rock, pop, and hip-hop genres. In this chapter we focus primarily on the chord progression. There are many variations on the blues progression, but the basic blues progression is shown in Figure 2.1.

Figure 2.1
Basic blues progression.

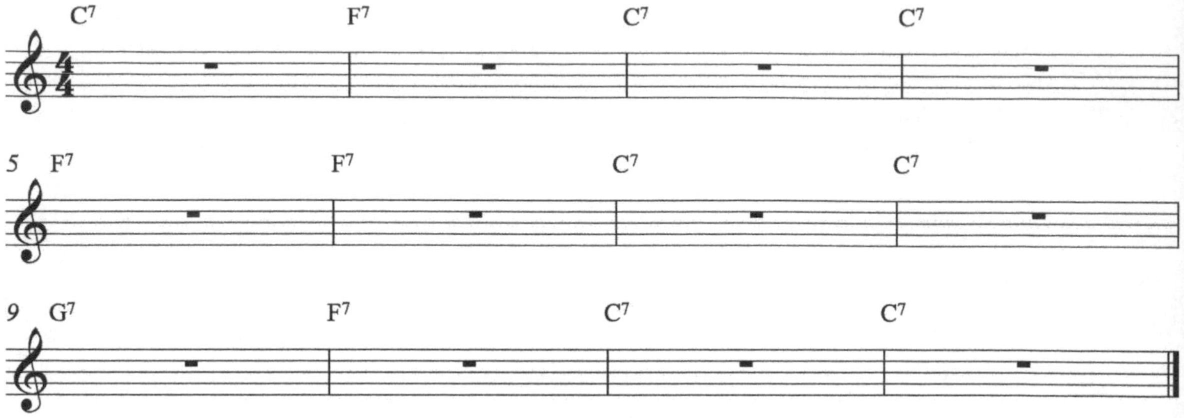

The basic blues is a 12-bar form and uses only three chords: I, IV, and V. All three chords are derived from the C major scale (see Figure 2.2).

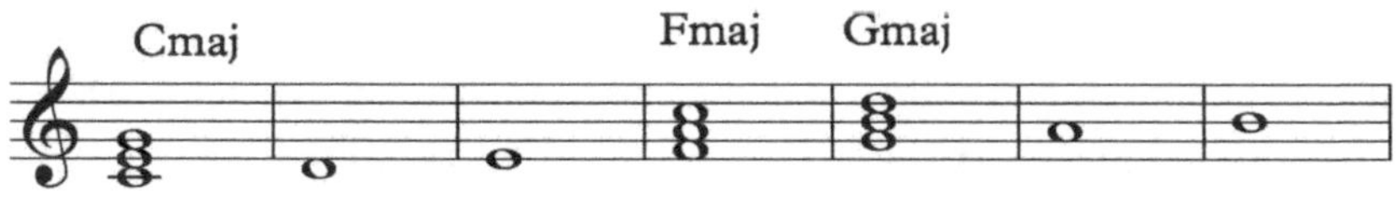

Figure 2.2
I, IV, and V chords derived from the C major scale.

Although it is possible to keep all of the chords as simple triads, it is common to add an interval of a minor 7th away from the root making the chords dominant and thus we have C^7, F^7, and G^7 (see Figure 2.3).

Figure 2.3
C^7, F^7, and G^7 chords.

When you are beginning to improvise on the blues progression, I suggest starting with the minor pentatonic scale based on the tonic of the song. So for a blues song in C, we will use the C minor pentatonic scale throughout the entire song. The minor pentatonic scale is constructed using 1 - ♭3 - 4 - 5 - ♭7. The C minor pentatonic scale is C - E♭ - F - G - B♭ (see Figure 2.4).

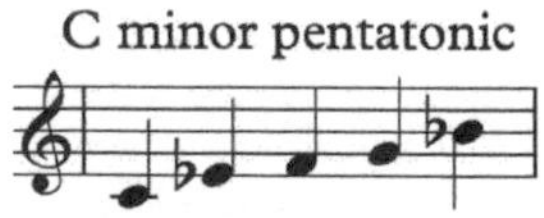

Figure 2.4
C minor pentatonic scale.

TIP

The minor pentatonic scale is also sometimes referred to as the 5th mode of the major pentatonic scale. For example, the E♭ major pentatonic scale (E♭ - F - G - B♭ - C) starting on the 5th scale degree (C) would yield the same notes of the C minor pentatonic scale described above.

Alternatively, one could also explore the blues scale based on the tonic of the song. The blues scale is essentially identical to the minor pentatonic with an added ♯4 (or ♯11). Think of the blues scale as the Swiss Army Knife of improvising. It is a scale that can work over many chords and provides a funky sort of feel—and I mean James Brown funky, not moldy cheese funky! To construct a blues scale, build the 1 - ♭3 - 4 - ♯4 - 5 - ♭7 from the root. So in a C blues, we'll play the C blues scale using C - E♭ - F - F♯ - G - B♭ - C (see Figure 2.5).

Figure 2.5
C blues scale.

Remember that the minor pentatonic scale and blues scales exists beyond the single octave written above. You should feel free to improvise using the minor pentatonic (see Figure 2.6) and blues scale (see Figure 2.7) notes above and below the given range. So extending the range of the scale above and below the single octave would give us the following:

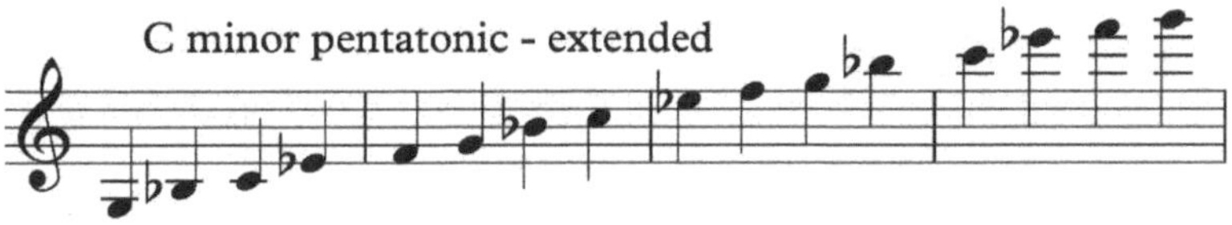

Figure 2.6
C minor pentatonic scale (extended range).

Figure 2.7
C blues scale (extended range).

If we compare the C7 chord to the minor pentatonic scale, it seems odd that a scale that uses an E♭ could work over a chord with an E♮. This brings me to the first rule of improvising:

If it sounds good, then it IS good. Trust your ear!

Our ears have grown up hearing a minor 3rd over a dominant chord and we have become used to the sound and, in fact, we usually really like that sound! This may not have sounded good to 19th-century ears, but it is perfectly normal to 21st-century ears.

Duke Ellington wrote "C Jam Blues," a simple 12-bar riff-based blues that follows the basic blues progression. The melody begins on the 5th and resolves up to the root (see Figure 2.8). This riff is played at the start of bar 1, 5, and 9 for a total of three times. This is one of my favorite beginner tunes because it uses a basic blues progression and I can teach the melody to anyone in minutes. I've actually taught this melody in scat-vocal style to three-year-olds and they were doo-dat-ting away with a Band-in-a-Box accompaniment and my saxophone in about five minutes!

Figure 2.8
Riff for the first 4 bars of "C Jam Blues."

A recording of the Lincoln Center Jazz Orchestra with Wynton Marsalis performing "C Jam Blues" can be found on the album *Live in Swing City—Swingin' with Duke* available on iTunes. You can also find many great recordings of "C Jam Blues" by doing a YouTube search. In Chapter 5 of this book (Web Resources for Listening) I discuss how to search YouTube and save various YouTube videos into a playlist for easy access and sharing with students. My favorite video comes from a 1942 video called "Jam Session" in which Duke slowly assembles his band in a local hangout and each member takes a one-chorus solo.

- YouTube Video Link: https://www.youtube.com/watch?v=gOlpcJhNyDI

TIP

A "chorus" means "once around the chord progression." So if someone takes a two-chorus solo over "C Jam Blues," they will play the entire progression twice for a total of 24 bars.

Although the key of C is usually a fairly friendly key, I recommend the key of E♭ for this tune. The reason for this choice will be more clear when we examine the corresponding minor pentatonic scale. So, since we're going to play "E♭ Jam Blues," the chord progression will look like Figure 2.9.

Figure 2.9

Chord progression for "C Jam Blues" in E♭.

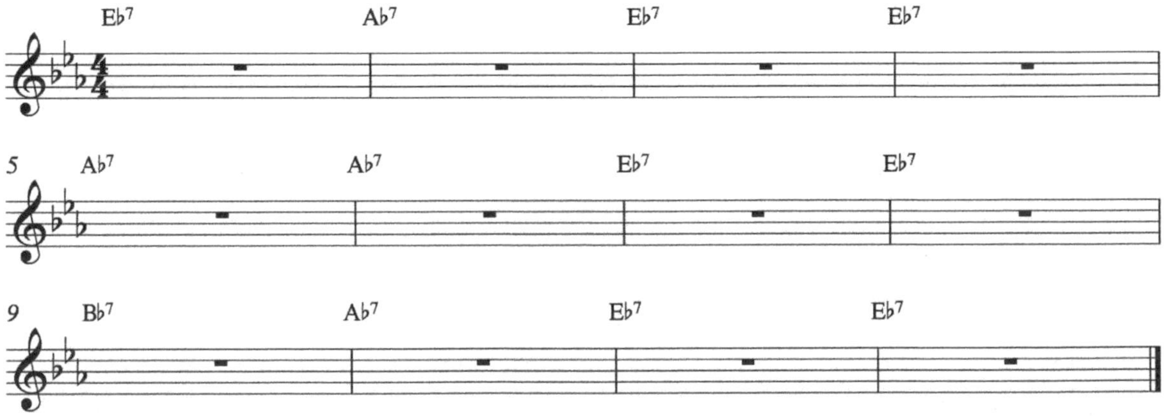

Figure 2.10

E♭ minor pentatonic scale.

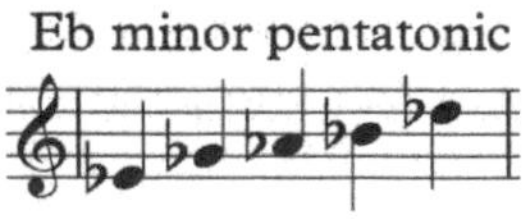

The corresponding minor pentatonic scale in E♭ will be E♭ - G♭ - A♭ - B♭ - D♭ - E♭ (see Figure 2.10). Examine these notes on a piano keyboard and you will notice that they are all black keys (see Figure 2.11). Isolating only black keys on the keyboard allows us to focus on phrasing and rhythm while the "right notes" will take care of themselves.

Figure 2.11

E♭ minor pentatonic scale on the black keys of the piano keyboard.

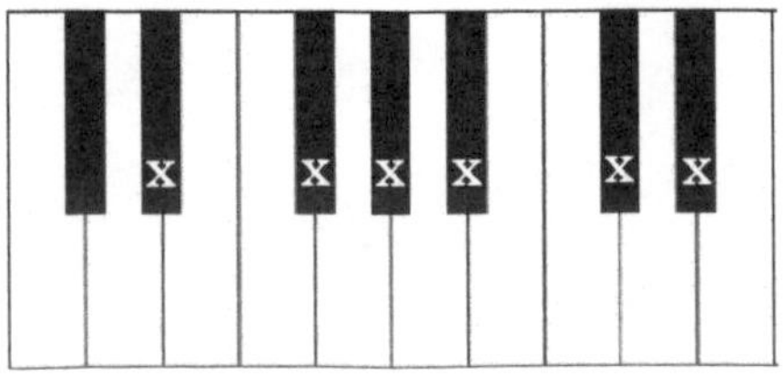

Remember, the E♭ blues scale works as well (E♭ - G♭ - A♭ - A♮ - B♭ - D♭ - E♭), so feel free to sprinkle in a ♯4 (A♮) here and there (see Figure 2.12).

The melody (see Figure 2.13) is also easier to locate since the 5th (B♭) sits at the top of the set of three black keys and the root (E♭) sits at the top of the set of two black keys on the piano keyboard (see Figure 2.14).

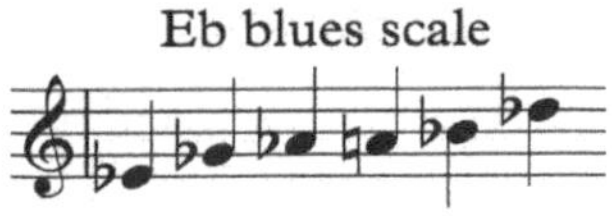

Figure 2.12
E♭ blues scale.

Figure 2.13
Riff for the first 4 bars of "C Jam Blues" in E♭.

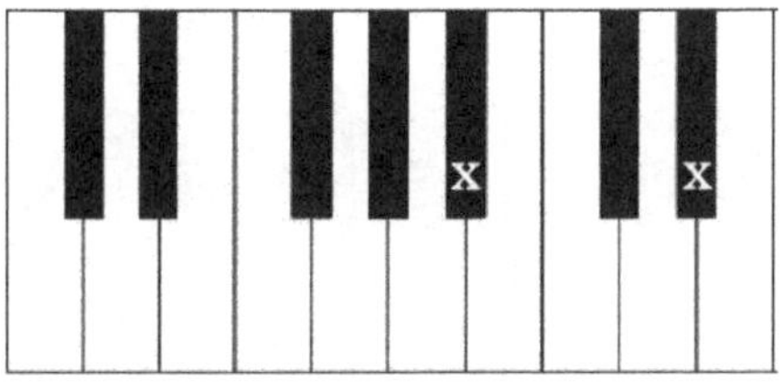

Figure 2.14
B♭ and E♭ on the piano keyboard. These are the only two notes for "C Jam Blues" in E♭.

Creating a Basic Accompaniment Track in Band-in-a-Box

[▶] *See Example 2.1 on the companion website for a Band-in-a-Box file and audio recording example of the "E♭ Jam Blues" activity.*

We are now going to create an accompaniment for "E♭ Jam Blues" in Band-in-a-Box. Although Band-in-a-Box has received many feature upgrades over the years, all of the activities we cover in Band-in-a-Box in this book should work just fine in older and newer versions alike, and most of the steps will be completely identical.

1. Launch Band-in-a-Box.
2. Click in the *Title* window. It will display "Untitled" when you open a new Band-in-a-Box file.
3. In the *Main Settings* dialogue box that pops up, enter the title (Eb Jam Blues), key (E♭), and a moderate tempo (120 bpm). See Figure 2.15.

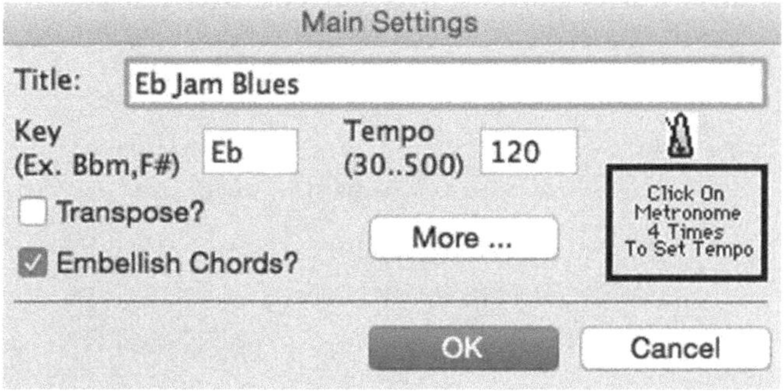

Figure 2.15
Main Settings dialogue box in Band-in-a-Box.

TIP

For most jazz accompaniments it is fine to keep "Embellish Chords" checked; however, I would uncheck this box for other genres such as classical/pop/rock/country accompaniments that typically have simpler chord structures with fewer chord alterations and extensions.

4. Click OK.
5. Set the length of the song by clicking on the box that says *End* and select 12.
6. Set the number of choruses to 4. This will allow us to perform the melody during choruses 1 and 4 and improvise during choruses 2 and 3

7. Click on the "C" in bar 1. Type "Eb7" on your QWERTY keyboard.

TIP

For flat symbols type lower-case "b" and for sharp symbols type the number sign (#).

TIP

Do not click the actual bar number. This will turn the color from blue (a) to green (b) to clear. Each Band-in-a-Box style includes two different feels, an A feel and a B feel. In an AABA song, we will set the bars so as to create a contrasting feel in the B section. In this song, we will keep the A feel only and Band-in-a-Box will automatically go to the B feel during the middle choruses. Band-in-a-Box does this because we often state the melody during the first and last chorus and improvise in the middle choruses.

8. Click the empty area next to bar 2 and type "Ab7."
9. Click the empty area next to bar 3 and type "Eb7."

TIP

Band-in-a-Box will continue to play this chord until there is a change so since bar 3 and bar 4 both use E♭7, you do not need to enter anything into bar 4.

10. Click the empty area next to bar 5 and type "Ab7."
11. Click the empty area next to bar 7 and type "Eb7."
12. Click the empty area next to bar 9 and type "Bb7."
13. Click the empty area next to bar 10 and type "Ab7."
14. Click the empty area next to bar 11 and type "Eb7." See Figure 2.16 for the full progression in Band-in-a-Box.

Figure 2.16

Chords for "E♭ Jam Blues" in Band-in-a-Box.

1a E♭7	2 A♭7	3 E♭7	4
5 A♭7	6	7 E♭7	8
9 B♭7	10 A♭7	11 E♭7	12

15. Press the PLAY button and listen to the accompaniment created by Band-in-a-Box. You should be hearing the default jazz style that includes piano, bass, and drums.

TIP

You can mute individual parts by right-clicking and selecting "Mute" in the pop-up window. So if you wanted to practice with only bass and drums, you could mute the piano part.

16. Although the standard jazz style is just fine, you may want to select a different style. Click on the *Style* menu at the top of the screen. At the bottom of this window you will see the 24 main styles that have been built into Band-in-a-Box since the 1990s. Go ahead and try out some of these various styles. Notice that most of the styles are in a meter of 4 (either 4/4 or 12/8) and only Jazz Waltz is in 3/4. If any style asks you to reduce or expand the chord duration almost always click "No."
17. Band-in-a-Box also comes with many other styles and, depending on the Band-in-a-Box package that you purchase, it may come with hundreds of other styles.
    a. To browse through the full set of styles, select Styles > Style Picker or simply click the Style icon located to left of the style info window under the title window.
    b. In the Style Picker (see Figure 2.17), select a category on the left pane and then a style in the large middle pane. Notice the description of each style on the right pane. It is very important to check the meter indications in the right pane to be sure you are selecting the appropriate meter for the song (in this case 4/4 or 12/8).

TIP

The meter in Band-in-a-Box is mostly tied to the style. You can, however, create mixed meter tunes but it is a bit labor intensive, especially if you have a lot of meter changes. To create a meter change, click and drag to select a measure or group of measures. Select *Edit > Set Time Sig of Scrap* and choose the appropriate time signature.

Figure 2.17

Style Picker dialogue box in Band-in-a-Box.

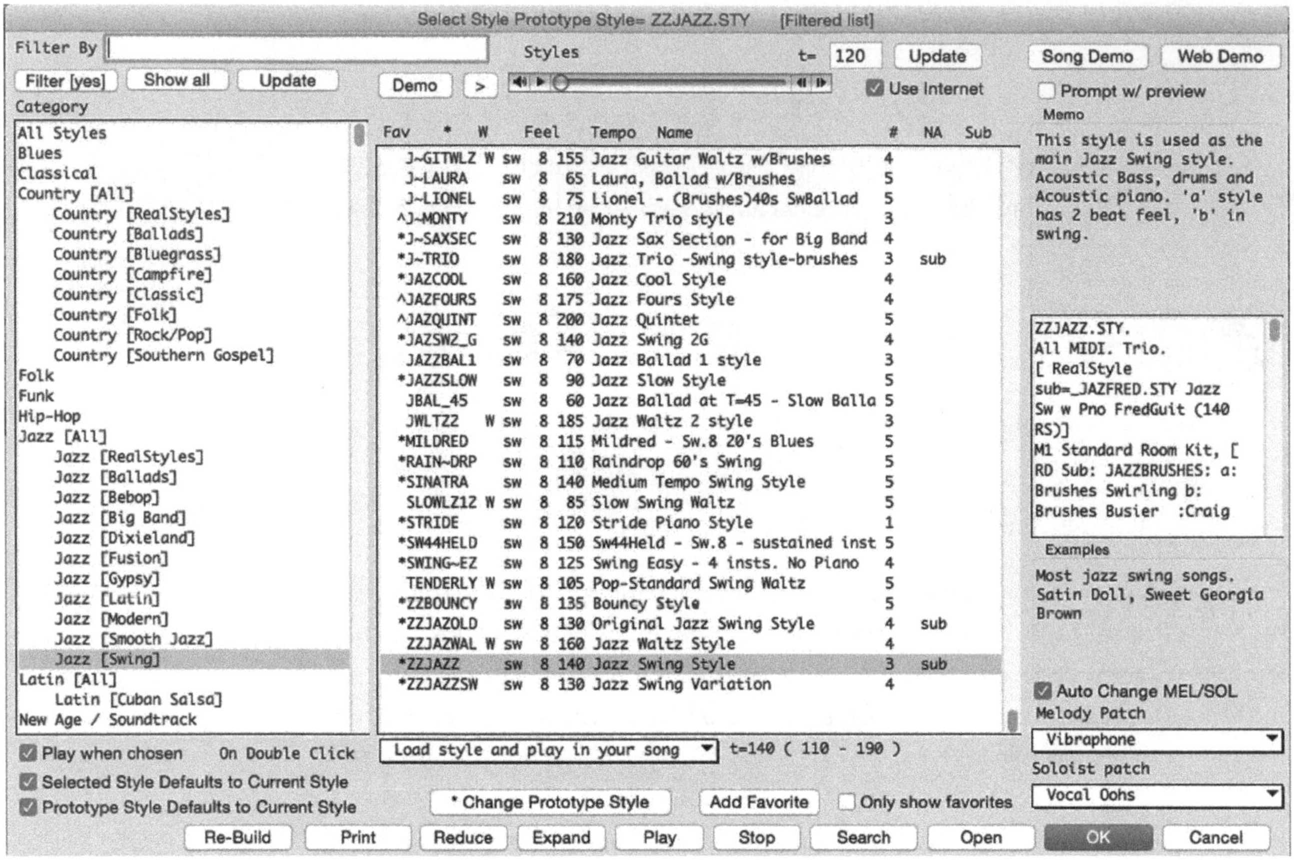

18. Now that you have a style, try performing "E♭ Jam Blues" on your piano keyboard along with the accompaniment.
19. If you have a MIDI keyboard connected to your computer, you can actually record in the melody. This is helpful for student practice.
    a. To set up your keyboard to work with Band-in-a-Box:
        i. *If you have a MIDI controller without General MIDI sounds:* Select MIDI > Choose Ports. See Figure 2.18 for the Port/Instruments window. Set the MIDI In to your MIDI controller. Set the Bass to Apple DLS Synth or any other synthesizer built in to your computer (such as the GS WaveTable on Windows). Click the button that says *Set all ports to this.*

Figure 2.18

Ports/Instruments window in Band-in-a-Box. Use these settings to have your computer produce the sound instead of your keyboard.

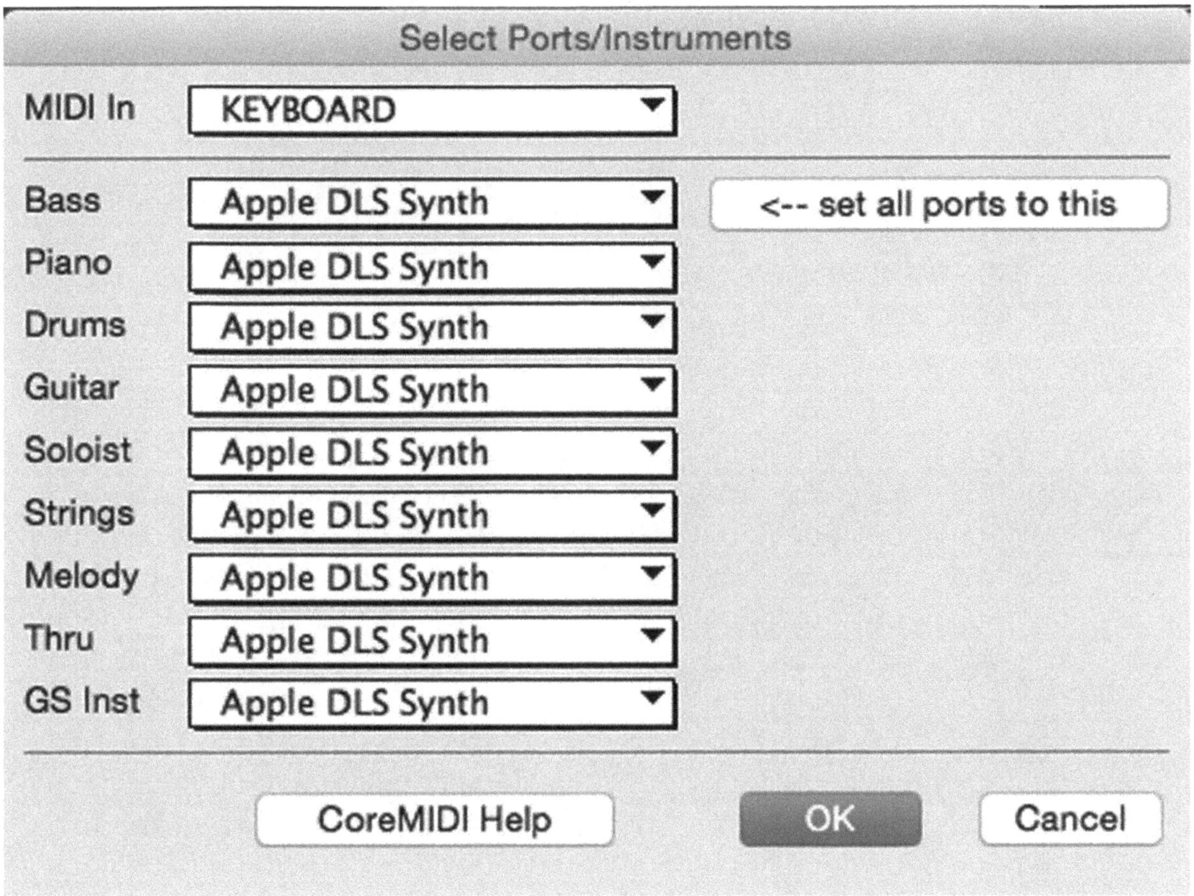

ii. *If you have a MIDI keyboard with General MIDI sounds:* Select MIDI > Choose Ports. Set the MIDI In to your MIDI keyboard/interface. Set the Bass to your MIDI keyboard/interface as well and click the button that says *Set all ports to this*. Set your MIDI keyboard into multi-timbral mode (sometimes called SEQUENCE mode or MULTI mode).

b. To begin recording, press the Record MIDI button and press *Record* in the dialogue box (see Figure 2.19). Perform the melody on your keyboard.

Figure 2.19
Record dialogue box in Band-in-a-Box.

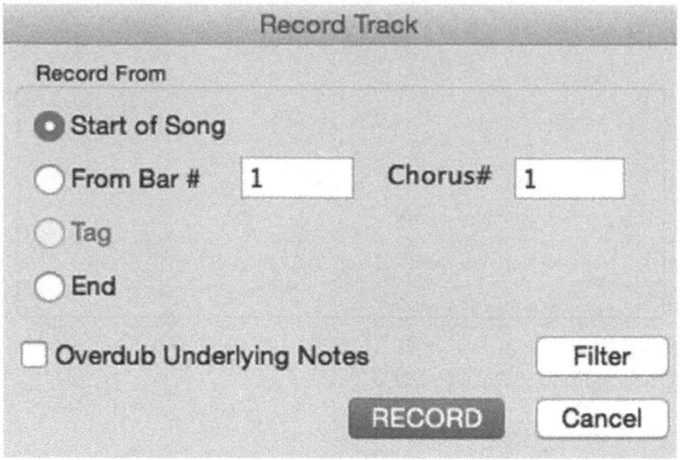

Figure 2.20
Melody Notes dialogue box in Band-in-a-Box.

c. Press the Spacebar to stop recording. In the pop-up dialogue box check the box for *Copy 1st Chorus to Whole Song* and click *OK - Keep Take* (see Figure 2.20).

d. To change the sound of the melody, right click *Melody* in the top and click *Select MIDI Patch > Select General MIDI Patch* (see Figure 2.21). Choose a GM sound in the list.

Figure 2.21
Right click *Melody* in the top and click *Select MIDI Patch > Select General MIDI Patch*.

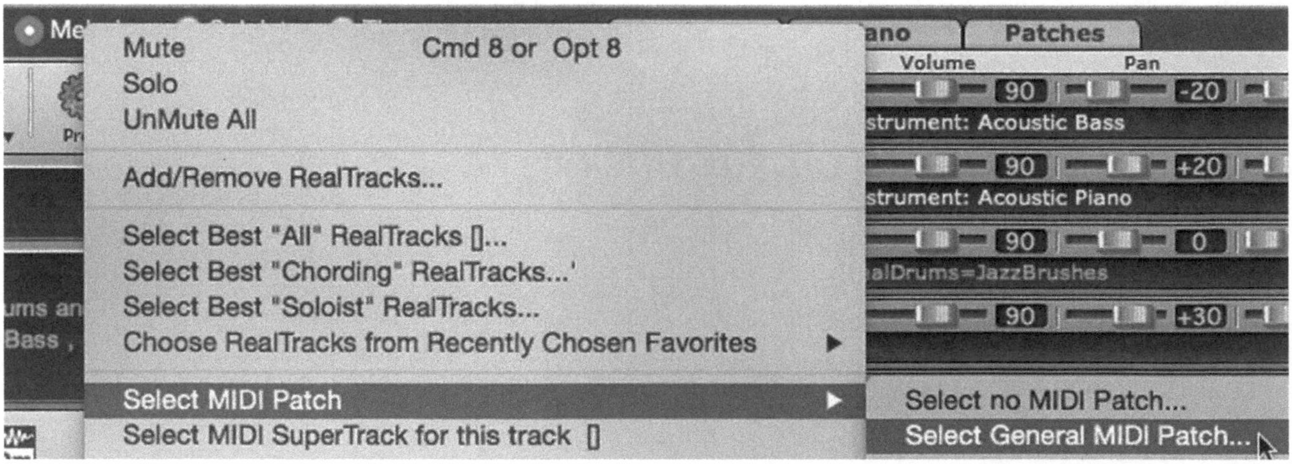

e. Listen to and evaluate your performance. Re-record if necessary.

20. The melody will typically only occur during the first and last choruses and your improvisation will typically occur during the middle choruses. To remove the melody during choruses 2 and 3, select *Melody > Kill Melody Choruses > Middle*

Choruses. Listen to your accompaniment and confirm that the melody has indeed been removed during the middle choruses.

21. Now we're ready to practice our improvisation.
    a. Review the E♭ minor pentatonic scale and practice playing the scale up and down at least two octaves on your piano keyboard (see Figure 2.22).

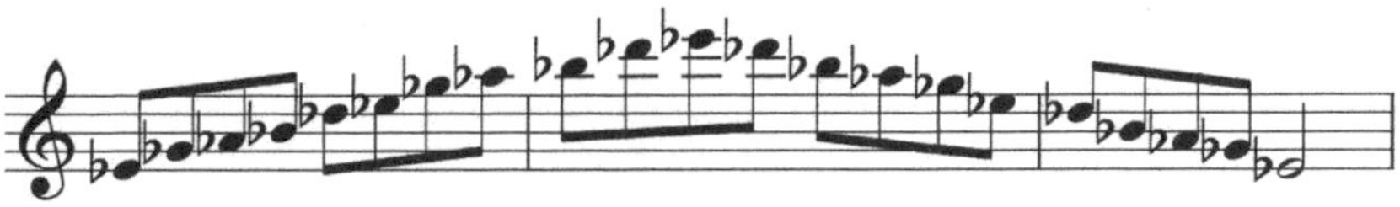

Figure 2.22
E♭ minor pentatonic scale two octaves.

    b. Practice playing the scale in 3rds (see Figure 2.23).

Figure 2.23
E♭ minor pentatonic scale in 3rds.

    c. Practice playing a returning scale pattern (1-2-3-1). See Figure 2.24 for this pattern.

Figure 2.24
E♭ minor pehtatonic scale in a returning scale pattern.

22. Try the various patterns indicated in the previous step while the Band-in-a-Box accompaniment plays. Listen to the sound of each note against the accompaniment as you play. Some notes like to sit while others have a strong sense of movement to another scale note. For example, on the E♭^{7} chord, the E♭, B♭, and D♭ notes love to just sit while the G♭ note craves movement to the E♭ and the A♭ note craves movement to the B♭.
23. Try improvising around the patterns in the previous steps. For example, try modifying the rhythm or dropping out a note.
24. Focus on phrasing. Remember that SPACE is just as important as your notes.
    a. Play a 1-bar phrase and then rest for 1 bar (see Figure 2.25). Repeat this throughout an entire chorus.

Figure 2.25

Play a 1-bar phrase and then rest for 1 bar.

b. Play a 2-bar phrase and then rest for 2 bars (see Figure 2.26). Repeat this throughout an entire chorus.

Figure 2.26

Play a 2-bar phrase and then rest for 2 bars.

c. Play a 3-bar phrase and rest for 1 bar (see Figure 2.27). Repeat this throughout an entire chorus.

Figure 2.27

Play a 3-bar phrase and rest for 1 bar.

d. Play a 2-bar phrase, rest for 1 bar, and then play a 1-bar phrase that leads into the next set of 4 bars (see Figure 2.28). For example, the phrase in bar 4 will lead into a new 2-bar phrase that begins in bar 5. Repeat this throughout an entire chorus. This is the most difficult phrasing idea to naturally feel and master. This sort of phrasing allows your improvised phrases to stretch over the bar. Phrasing in this manner will give your improvisation a feeling of movement as opposed the other phrasing structures that tend to make your improvisation feel more grounded.

Figure 2.28

Play a 2-bar phrase, rest for 1 bar, and then play a 1-bar phrase that leads into the next set of 4 bars.

25. Save your Band-in-a-Box file to a memorable location.

Blues Variation: "Freddie Freeloader"

Miles Davis released the album *Kind of Blue* in 1959. In stark contrast to the increasingly complex melodic and harmonic structure of bebop and hard bop of the '40s and '50s, *Kind of Blue* includes five tunes with simple melodies and chord progressions perfect for beginning and experienced improvisers alike. When improvising on these tunes, the player generally focuses on a scale or mode that corresponds to the given chord. The improviser uses the mode to create melodic phrases remembering to change modes when the chord changes. This album is one of the best- (if not THE best-) selling jazz album of all time and it still sells more than 5,000 copies a week. After one listen you'll know why. Dan Morgenstern, acclaimed jazz writer and critic, explained, "It doesn't wear out its welcome."[1] The simplicity and balance bring the listener back again and again, finding new things to enjoy each time.

Each musician on the album brings a very unique style and the blend of the group provides a powerful impression. Focusing on the horn players, Miles Davis (trumpet) plays simple phrases, almost always within the mode, with a vulnerable tone. John Coltrane (tenor saxophone) enters with a powerful sound and sheets of notes up and down. Julian Cannonball Adderley (alto saxophone) uses interesting out-of-mode notes while maintaining his hard, driving sound and perfectly swinging rhythms. Each player approaches the simple canvas of the basic chord progression and melody differently and reminds us how improvisation really allows the individual musician to tell his or her story.

Explore the full article about *Kind of Blue* from NPR.org:

http://www.npr.org/programs/jazzprofiles/archive/miles_kob.html.

Listen to *Kind of Blue* on iTunes:

https://itunes.apple.com/us/album/kind-of-blue/id268443092.

You can find sheet music for "Freddie Freeloader" in *The Real Book* (6th ed.), Volume I, published by Hal Leonard,

http://www.halleonard.com/product/viewproduct.action?itemid=240221&.

"Freddie Freeloader" is essentially a basic blues in B♭ with a twist. Instead of moving to the IV^7 chord in bar 2, Miles stays on the I^7 chord for a full 4 bars. In the last 4-bar phrase, the movement of the V^7 to the IV^7 doesn't resolve to the I^7 and instead resolves to ♭VII^7 chord, certainly a unique choice. The chord progression is as follows (see Figure 2.29):

Figure 2.29

"Freddie Freeloader" chord progression.

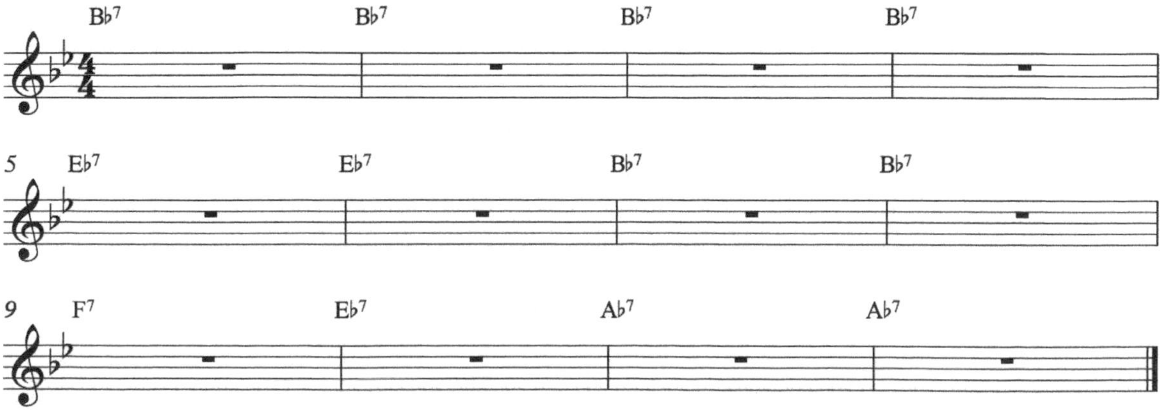

1 Jim Luce, "Jazz Profiles from NPR: Miles Davis: Kind of Blue," *NPR.org*, accessed April, 11, 2016. http://www.npr.org/programs/jazzprofiles/archive/miles_kob.html.

Although we could certainly use the B♭ minor pentatonic scale and/or blues scale over this entire tune, we will focus more on the modes (and chord tones) for each chord to create the modal sound featured on *Kind of Blue*.

First let's write out an E♭ major scale—yes, E♭, not B♭. This E♭ major scale is a parent scale because seven modes and chords come from it. If we build a scale or mode on each note of the E♭ major scale, we get the following scales (see Figure 2.30):

Figure 2.30

E♭ major scale modes.

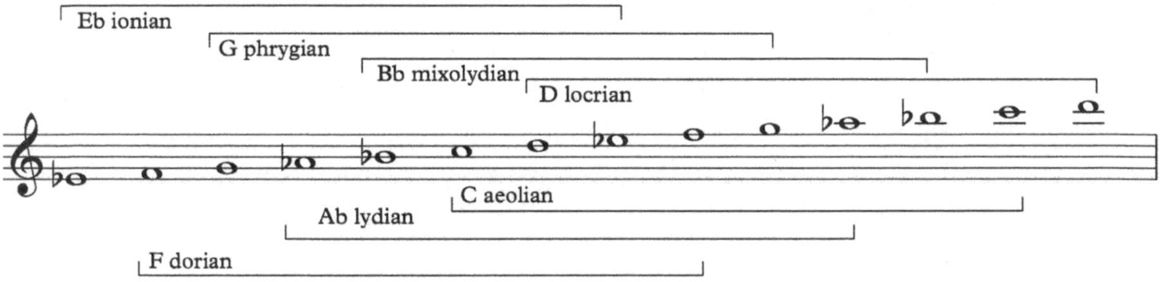

If we create 7th chords on each scale degree we get the chords in Figure 2.31.

Figure 2.31

7th chords derived from the E♭ major parent scale.

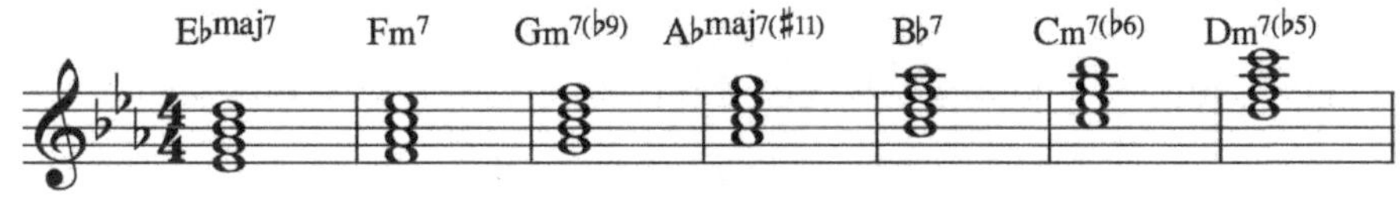

There is a direct correlation between the chords and scales. You can play the E♭ major/ionian scale over E♭Maj^{7}, F dorian scale over Fmin^{7}, and so forth.

Chord	**Scale**
E♭Maj^{7}	E♭ ionian (major)
Fmin^{7}	F dorian
Gmin$^{7(\flat 9)}$	G phrygian
A♭Maj$^{7(\sharp 11)}$	A♭ lydian
B♭^{7}	B♭ mixolydian
Cmin$^{7(\flat 6)}$	C aeolian (natural minor)
Dmin$^{7(\flat 5)}$ or D^{ø7}	D locrian

TIP

The ii, iii, and vi chords will typically be notated simply as a min7 chord without an alteration. I discuss selecting the appropriate mode for min7 chords in the section *Modal Improvisation: So What* presented later in this chapter.

In this tune, we have a $B\flat^7$ chord and thus the appropriate mode for that chord is the B♭ mixolydian scale (parent scale: E♭ major). So extending this concept to the other chords would provide that $E\flat^7$ uses an E♭ mixolydian scale (parent scale: A♭ major), F^7 uses an F mixolydian scale (parent scale: B♭ major), and the $A\flat^7$ uses an A♭ mixolydian scale (parent scale: D♭ major). See Figure 2.32 for each of these scales notated on the treble cleff staff.

Figure 2.32

Mixolydian modes needed for "Freddie Freeloader."

TIP

For younger musicians, I typically will write in the appropriate parent scale over the chord change. So for example, I'd write in "E♭ major scale" over the $B\flat^7$ chord. For more advanced musicians, I will typically write in the appropriate mode. So in the same example, I'd write "B♭ mixolydian" over the $B\flat^7$ chord.

TIP

I would also suggest that advanced students think less about the parent scale and more about how the mode modifies the root note major scale. For example, when I see $B\flat^7$, I don't usually think about E♭ major. Instead, I know that mixolydian adds a flat 7, so I improvise on the B♭ major scale with a flat 7 (A♭ in this case). I find this a quicker way for the brain to access the proper notes and make the connection between the chord tones and the scale. Here's a chart describing how the mode modifies the root note major scale for all major scale modes:

Mode	Root Note Major Modification
Ionian	No changes - same as major
Dorian	♭3, ♭7
Phrygian	♭2, ♭3, ♭6, ♭7
Lydian	♯4
Mixolydian	♭7
Aeolian	♭3, ♭6, ♭7
Locrian	♭2, ♭3, ♭5, ♭6, ♭7

Creating an Accompaniment Track for "Freddie Freeloader"

[▶] *See Example 2.2 on the companion website for a Band-in-a-Box file and audio recording example of the "Freddie Freeloader" activity.*

Let's create an accompaniment in Band-in-a-Box so we can practice the mixolydian modes on the chords to "Freddie Freeloader."

1. Launch Band-in-a-Box.
2. Click in the *Title* window. It will display "Untitled" when you open a new Band-in-a-Box file.
3. In the *Main Settings* dialogue box that pops up, enter the title (Freddie Freeloader), key (Bb), and a moderately slow tempo (120 bpm). It is fine to keep *Embellish Chords* checked for this song.
4. Click OK.
5. Set the length of the song by clicking on the box that says *End* and select 12.
6. Set the number of choruses to 5. This will allow us to perform the melody during choruses 1 and 5 and improvise during choruses 2–4.
7. Click on the "C" in bar 1. Type "B♭7" on your QWERTY keyboard. Remember, Band-in-a-Box will continue to play this chord until there is a change; since bars 1–4 all use B♭7, you do not need to enter anything into bars 2–4.
8. Click the empty area next to bar 5 and type "Eb7."
9. Click the empty area next to bar 7 and type "Bb7."
10. Click the empty area next to bar 9 and type "F7."
11. Click the empty area next to bar 10 and type "Eb7."
12. Click the empty area next to bar 11 and type "Ab7." See Figure 2.33 for a screenshot of the full chord progression entered into Band-in-a-Box.

Figure 2.33
"Freddie Freeloader" chord progression entered in Band-in-a-Box.

13. Press the PLAY button and listen to the accompaniment created by Band-in-a-Box. You should be hearing the default jazz style that includes piano, bass, and drums.
14. Try selecting a different style.
    a. Click on the Style menu at the top of the screen. Try out the 24 main styles listed at the bottom of that menu. Remember, if any style asks you to reduce or expand the chord duration, almost always click *No*.
    b. Browse through the full set of styles by clicking the Style icon located to left of the style info window under the title window. In the Style Picker,

select a category on the left pane and then a style in the large middle pane. You can select any genre; don't feel that you need to stay in the *Jazz* genre simply because the tune was originally a jazz tune. Also be aware of the meter of the style you are selecting. I would recommend staying in 4/4 or 12/8 but you can feel free to explore styles with other meters such as 3/4 or 5/4.

15. If you have a MIDI keyboard connected to your computer, record in the melody. Listen to and evaluate your performance. Re-record if necessary. Once you have a good take, check the box for *Copy 1st Chorus to Whole Song* and click *OK - Keep Take.*
16. Remove the melody during choruses 2–4 by selecting *Melody > Kill Melody Choruses > Middle Choruses.* Listen to your accompaniment and confirm that the melody has indeed been removed during the middle choruses.
17. Now we're ready to practice our improvisation. The goal is to perform the appropriate mixolydian mode over each chord. Confine yourself to only playing notes from the mode. It is a good idea to slow down the tempo quite a bit to get comfortable with the scales. Start with 60 bpm or slower to get started.
    a. Practice each mode on its own. I suggest running eighth notes up and down the scale once.
    b. Practice going from mode to mode. Meaning, play the mixolydian scales in the order of the song (B♭ mixolydian to E♭ mixolydian to B♭ mixolydian to F mixolydian to E♭ mixolydian to A♭ mixolydian and finally back B♭ mixolydian). See Figure 2.34 for an example of this exercise.

Figure 2.34

Play the mixolydian scales in the order of "Freddie Freeloader."

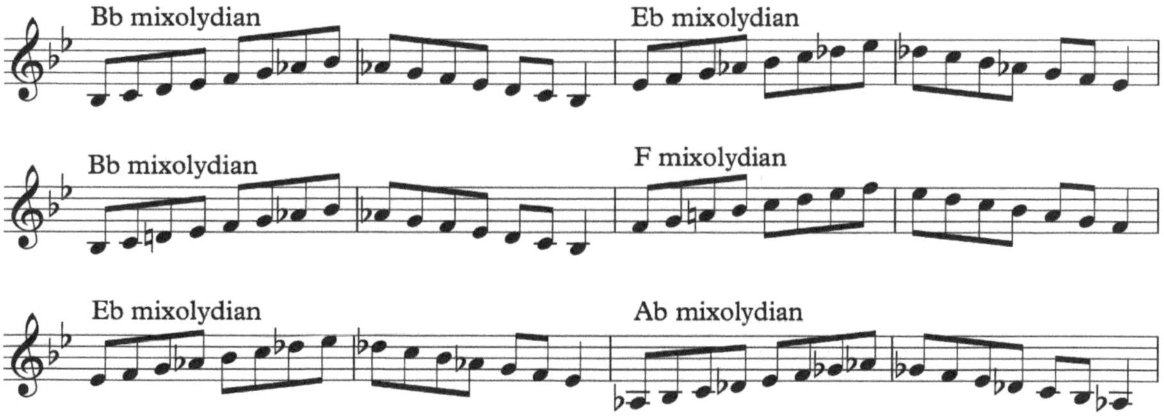

    c. Practice playing the modes using eighth notes up and down the scale along with the accompaniment at a slow tempo. Slowly speed up the tempo with a goal of approximately 140 bpm.
    d. Practice playing various permutations of the mode such as the scale in 3rds along with the accompaniment.

e. Compose a short 1-bar melodic motif and perform it in B♭, E♭, F, and A♭ along with the accompaniment. You may want to write out the patterns (software notation is discussed in Chapter 3). See Figure 2.35 for an example of a 1-bar melodic motif transposed to each required key for this song.

Figure 2.35

1-bar melodic motif transposed to each required key for "Freddie Freeloader."

18. Once you are comfortable playing specific patterns, try improvising around the patterns in the previous steps. For example, try modifying the rhythm or dropping out a note.
    a. Figure 2.36 shows some examples of permutations of the pattern in the previous step.

Figure 2.36

Permutations of the pattern in Figure 2.35.

19. Don't forget to focus on phrasing. Leave space for the listener to digest your musical ideas.
    a. Play a 1-bar phrase and then rest for 1 bar (see Figure 2.37). Repeat this throughout an entire chorus.

Figure 2.37

Play a 1-bar phrase and then rest for 1 bar.

b. Play a 2-bar phrase and then rest for 2 bars (see Figure 2.38). Repeat this throughout an entire chorus.

Figure 2.38

Play a 2-bar phrase and then rest for 2 bars.

c. Play a 3-bar phrase and rest for 1 bar (see Figure 2.39). Repeat this throughout an entire chorus.

Figure 2.39

Play a 3-bar phrase and rest for 1 bar.

d. Play a 2-bar phrase, rest for 1 bar, and then play a 1-bar phrase that leads into the next set of 4 bars (see Figure 2.40). Repeat this throughout an entire chorus.

Figure 2.40

Play a 2-bar phrase, rest for 1 bar, and then play a 1-bar phrase that leads into the next set of 4 bars.

TIP

If you are improvising for an extended amount of time (two or more choruses) think about building your improvisation from simple to complex. The easiest way to achieve this is to start the first chorus with more space and long duration notes. By the time you reach the final chorus it would be normal to have less space and more eighth and sixteenth notes.

20. Record a take of your improvised solo using Band-in-a-Box.
    a. Select *Soloist > Edit Soloist Track > Record Soloist* (see Figure 2.41).

Figure 2.41

Select *Soloist > Edit Soloist Track > Record Soloist.*

   b. Click the button next to *From Bar #* and enter "2" into the *Chorus #* box.
   c. Press *Record*. Band-in-a-Box will give you a 2-bar lead-in.
   d. Perform an improvised solo.
   e. Press ESC to stop the recording and then press *OK—Keep Take* in the pop-up dialogue box.

21. View the notation of your solo.
   a. Click the *Notation* icon.
   b. Click the "S" for soloist in the instrument list. B D P G S M S

TIP

The first "S" in the list is for strings. The second "S" is for soloist.

22. Press play to listen to your improvisation.
   a. Click the *Play from Bar* icon.
   b. Click measure 1 and select chorus #2 from the pop-up dialogue.
23. Evaluate your performance based on following criteria:
   a. Did you restrict yourself to only the notes from the mixolydian mode?
   b. Did you change modes at the appropriate time?
   c. Did you leave SPACE?
   d. Did you follow the suggested phrasing models?
   e. Did your improvisation start simple in chorus 1 and slowly increase in complexity during chorus 2 and chorus 3?

TIP

Don't worry if you didn't execute everything perfectly. Improvisation is a marathon, not a sprint, and it will take time for you to get comfortable. Celebrate your successes and identify your weaknesses. If you were only successful remembering to leave space, I'd consider that a victory!

24. Save your Band-in-a-Box file to a memorable location.

Enrichment: Shell Voicing (3rd/7th)

The 3rd and 7th of each chord are essential to determining the chord quality. Often these notes are referred to as guide tones because they can help guide a player through the progression, often alternating between the 3rd and the 7th. For example, let's look at the first three chords in the basic blues and how guide tones can improve the voice-leading between each.

Here's the root voicing of the basic blues progression (see Figure 2.42).

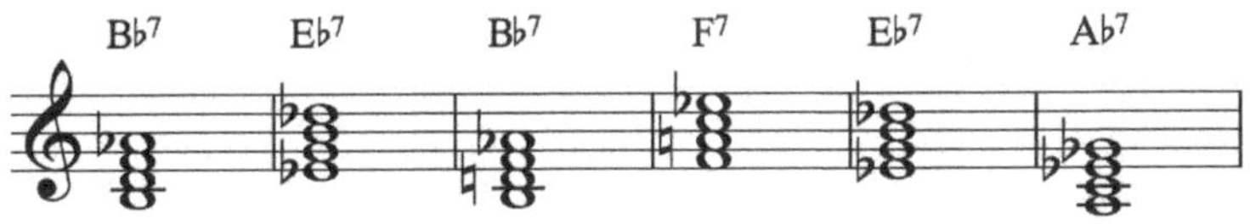

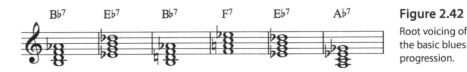

Figure 2.42 Root voicing of the basic blues progression.

Assuming the bass player or the piano left hand is playing the root or root/5th, we can focus on the 3rd and 7th voicing in the treble staff. In an B♭7 chord, the 3rd is D and the 7th is A♭. In an E♭7 chord, the 3rd is G and the 7th is D♭. To improve the voice leading between chords, voice the B♭^{7} with D and A♭ and the E♭^{7} with D♭ and G. You can apply similar principles to the F^{7} chord using E♭ and A and the A♭^{7} chord using C and G♭. See the example voicing below in Figure 2.43.

Figure 2.43 Shell voicing for "Freddie Freeloader."

Perform this voicing throughout the basic blues progression. Improvise the rhythm to add interest.

You can also try flipping this voicing so the B♭^{7} chord is voiced with A♭ and D (see Figure 2.44).

Figure 2.44 Alternate shell voicing for "Freddie Freeloader."

Perform this voicing throughout the progression. Experiment with the following rhythms while you "comp" the chords. See Figures 2.45, 2.46, and 2.47 for example rhythms.

Perform this voicing and improvise the rhythm. You can integrate the rhythms along with your own ideas.

Figure 2.45
Comping rhythm #1.

Figure 2.46
Comping rhythm #2.

Figure 2.47
Comping rhythm #3.

Modal Improvisation: "So What"

"So What" is the first tune on Miles Davis's *Kind of Blue*. The bass actually plays the melody with simple piano and horn hits accompanying. The tune is 32 bars in an AABA form with the A sections in D minor and the B section in E♭ minor. The A and B sections are completely identical except for the half step transposition. The chord progression for "So What" is shown in Figure 2.48.

Figure 2.48
Chord progression for "So What."

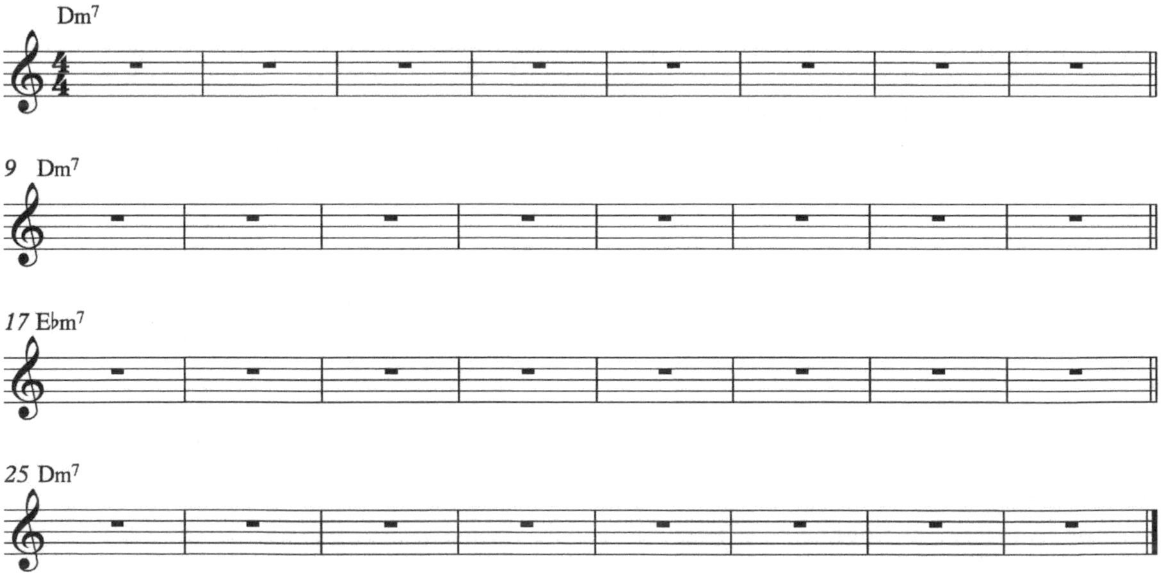

You can find sheet music for "So What" in *The Real Book* (6th ed.), Volume I, published by Hal Leonard.

http://www.halleonard.com/product/viewproduct.action?itemid=240221&

TIP

The chord progression for "So What" has also served as the basis for another popular tune by John Coltrane called "Impressions." In contrast to "So What," "Impressions" is generally played at a very fast tempo and includes a more technically challenging melody played by the horn player. Interestingly enough, the melody for "Impressions" actually came from *Pavanne, Mvmt. 2*, from Morton Gould's *American Symphonette No. 2*, a well-known classical piece composed in 1939.

For the full story of how John Coltrane combined Gould's *Pavanne* and Miles Davis's "So What" to create "Impressions," check out this article by Dr. Lewis Porter:

http://www.wbgo.org/blog/dr-lewis-porter-on-john-coltrane-impressions-part-two.

Similarly to "Freddie Freeloader," we will again focus on the appropriate mode for each chord. We examined the various chords derived from the major parent scale previously and we found that minor chords are found on the ii, iii, and iv scale degrees. "So What" uses the ii minor chord, and therefore the dorian scale is most appropriate.

TIP

How do you know whether the minor chord is derived from the ii, iii, vi, or something different all together? Well, that answer really depends on the specific placement of the chord, the chords around it, and the melody of the tune. In this case, the horn/piano hits use the notes from the dorian scale and that helps us determine that dorian is most appropriate for improvising.

As previously mentioned, you won't typically find a chord notated as minor$^{7\flat 9}$ or minor$^{7\flat 6}$ and instead you will only find chords notated as min7 without any alteration. You must rely on **context** to select the appropriate mode. For example, if you have a progression such as Emin7-Amin7-Dmin7-G7-CMaj7, I would play E phrygian over Emin7 (functioning as the iii), A aeolian over Amin7 (functioning as the vi), and D dorian over Dmin7 (functioning as the ii) since the context of the chords places them all in the key of C major. In contrast, if you had an Emin7 resolving to DMaj7, I would play E dorian over Emin7 since this chord is functioning as the ii in the key of D major.

It is also advisable to examine the melody notes to inform your choice as illustrated in "So What." All of that said, in the end, let your ears be the judge.

In this tune, we encounter both D minor and E♭ minor. Since dorian is based on the second scale degree of the major parent scale, D dorian comes from C major (see Figure 2.49) and E♭ dorian comes from D♭ major (see Figure 2.50).

Figure 2.49

D dorian scale derived from C major scale.

Figure 2.50
E♭ dorian scale derived from D♭ major scale.

TIP

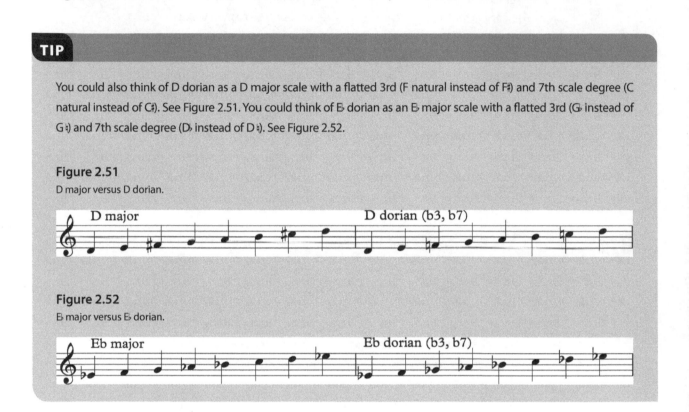

You could also think of D dorian as a D major scale with a flatted 3rd (F natural instead of F♯) and 7th scale degree (C natural instead of C♯). See Figure 2.51. You could think of E♭ dorian as an E♭ major scale with a flatted 3rd (G♭ instead of G♮) and 7th scale degree (D♭ instead of D♮). See Figure 2.52.

Figure 2.51
D major versus D dorian.

Figure 2.52
E♭ major versus E♭ dorian.

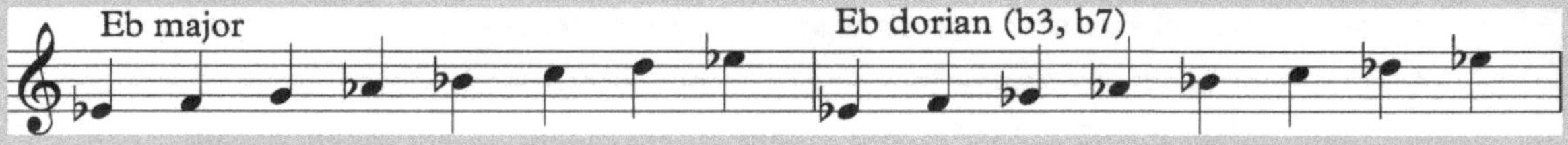

Creating an Accompaniment Track for "So What"

[▶] *See Example 2.3 on the companion website for a Band-in-a-Box file and audio recording example of the "So What" activity.*

Let's create an accompaniment in Band-in-a-Box for "So What." Since this tune only changes chords twice, you will have an accompaniment ready in minutes!

1. Launch Band-in-a-Box.
2. Click in the *Title* window. It will display "Untitled" when you open a new Band-in-a-Box file.
3. In the *Main Settings* dialogue box that pops up, enter the title (So What), key (Dm), and a moderately slow tempo (120 bpm). It is fine to keep *Embellish Chords* checked for this song.
4. Click OK.
5. The length of the song should already be set to 32 since this is the default value. If it is not set to the 32, click on the box that says *End* and select 32.
6. Set the number of choruses to 5. This will allow us to perform the melody during choruses 1 and 5 and improvise during choruses 2–4.

7. Click on the "C" in bar 1. Type "Dm7" on your QWERTY keyboard.
8. Click the empty area next to bar 17 and type "Ebm7."
9. Click the empty area next to bar 25 and type "Dm7."
10. To change the style to the "B" feel, click the actual number at bar 17 twice so it turns green and says 17b.
11. Click the actual number at bar 25 once so it turns blue and says 25a. See Figure 2.53 for a screenshot of the chord progression entered into Band-in-a-Box.

Figure 2.53

Chord progression for "So What" entered into Band-in-a-Box.

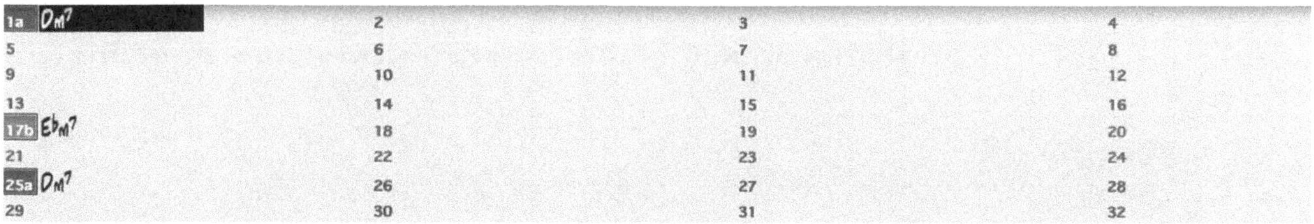

12. Play the accompaniment. You should hear a change at bar 17 and a return to the original feel at bar 25. Remember, Band-in-a-Box defaults to the "B" feel throughout all middle choruses.
13. Try selecting various styles by clicking the Style icon. 
14. If you have a MIDI keyboard connected to your computer, record in the piano/horn part as the melody of the song. See Figure 2.54 for notation of the horn hits in both keys. We won't record the actual melody played by the bass, but feel free to learn and perform that part along with the accompaniment and your newly recorded hits. Listen to and evaluate your performance. Re-record if necessary.

Figure 2.54

"So What" horn hits.

15. Once you have a good take, check the box for *Copy 1st Chorus to Whole Song* and click *OK - Keep Take.*
    a. In a few steps, we will use a feature called the Wizard. The Wizard records into the melody track, so for now, we need to move the melody you just recorded to the soloist track. To do this select *Melody > Swap Melody and Soloist Track.* After we have finished with the Wizard feature, we can swap them back.
16. Now we're ready to practice our improvisation. The goal is to perform the appropriate dorian mode over each chord. Confine yourself to only playing notes from the mode. It is a good idea to slow down the tempo to get comfortable with the scales.

a. Practice D dorian and E♭ dorian on its own. I suggest running eighth notes up and down the scale once (see Figure 2.55).

Figure 2.55

D dorian and E♭ dorian in eighth notes up and down the scale.

b. Practice going back and forth from D dorian and E♭ dorian to help your brain make the change and fluidly go from one to the other and back. You can also try ascending on D dorian and then descending on E♭ dorian and vice versa (see Figure 2.56).

Figure 2.56

Ascending on D dorian and then descending on E♭ dorian and vice versa.

c. Practice playing the modes using eighth notes up and down the scale along with the accompaniment at a slow tempo. Slowly speed up the tempo with a goal of approximately 140 bpm.
d. Practice playing various permutations of the mode such as the scale in 3rds along with the accompaniment (see Figure 2.57).

Figure 2.57

D dorian and E♭ dorian in 3rds.

e. Compose a short 1-bar melodic motif and perform it in D dorian and E♭ dorian along with the accompaniment. You may want to write out the patterns (software notation is discussed in Chapter 3). See Figure 2.58 for an example motif in D and E♭.

Figure 2.58
1-bar melodic motif in D dorian and E♭ dorian.

17. Once you are comfortable playing specific patterns, try improvising around the patterns in the previous steps. For example, try modifying the rhythm or dropping out a note.
18. Don't forget to focus on phrasing. Leave space for the listener to digest your musical ideas.
    a. In a tune like "So What" it is very effective to begin a phrase 1 or 2 bars before the change from D minor to E♭ minor or from E♭ minor to D minor. This helps connect the chords and give the listener a feel of *moving to* a new section instead of just starting in the new section. See the notation in Figures 2.59 and 2.60 below for two examples.

Figure 2.59
Example of phrasing from D dorian into E♭ dorian.

Figure 2.60
Example of phrasing from E♭ dorian into D dorian.

19. At this point, you could go in two different directions. As we did with "Freddie Freeloader," it would be great to record a take of your improvised solo using Band-in-a-Box, view the notation of your solo, and evaluate your performance. But for this song, I'd like to explore one of the most unique features in Band-in-a-Box called the Wizard. This feature turns your QWERTY keyboard into a performance device. The bottom row of keys (Z to the backslash symbol) will play chord tones while the second row of keys (A to the apostrophe symbol) will play non-chord tones.

Non-Chord Tones:	a	s	d	f	g	h	j	k	l	;	'
Chord Tones:	z	x	c	v	b	n	m	,	.	/	

    a. Select the *Song* menu at the top of the screen and be sure to enable the *Wizard Playalong*.
    b. Press play and play the bottom row of keys. Notice that the keys change notes when you get to the E♭ minor. Cool!

TIP

Press the SHIFT key to go up an octave while playing with the Wizard feature.

   c. Now try the non-chord tones (second row of keys). If you only play these notes, the improvisation won't fit the chords very well. Instead, try to play 1–3 chord tones followed by a non-chord tone and resolving to a chord tone.
   d. Try playing only the bottom row keys for chorus 1 and then incorporate the second row keys during the next chorus to add interest.
   e. Be sure to focus on the phrasing ideas mentioned previously to create a sensical improvisation.
   f. Once you feel comfortable with the Wizard, press record. The Wizard will record into the melody track.
   g. Select *Melody > Swap Melody and Solist Track* to place your MIDI recordings in the proper tracks.

20. View the notation of your solo.
    a. Click the *Notation* icon. 
    b. Click the "S" for soloist. 

21. Press play to listen to your improvisation.
22. Evaluate your performance based on the following criteria:
    a. Did you leave SPACE?
    b. Did you begin phrases 1 or 2 bars before a chord change to lead into that change and connect the chords?
    c. Did your improvisation start simple in the chorus 1 (mostly chord tones) and slowly increase in complexity during the following choruses (including more non-chord tones)?
    d. Did you resolve non-chord tones to chord-tones?
23. Save your Band-in-a-Box file to a memorable location.

Chord Bracketing: "Autumn Leaves"

"Autumn Leaves" was originally composed in French by Joseph Kosma and Jacques Prevert in 1945. Johnny Mercer wrote English lyrics two years later and the tune has been a jazz and pop hit ever since, recorded by Frank Sinatra, Miles Davis, Bill Evans, Chet Baker, Andrea Bocelli (French version), Eric Clapton, Bob Dylan, and many, many more. My personal favorite versions are by Cannonball Adderely and Miles Davis on *Somethin' Else* and Sonny Stitt and Gene Ammons on *Boss Tenors*.

Listen to "Autumn Leaves" as played by Cannonball Adderley and Miles Davis on iTunes:

https://itunes.apple.com/mt/album/cannonball-adderley-autumn/id591130536.

Listen to "Autumn Leaves" as played by Sonny Stitt and Gene Ammons on iTunes:

https://itunes.apple.com/us/album/boss-tenors/id534821238.

"Autumn Leaves" is a great beginning tune because, although there are chord changes in just about every bar, the entire tune has only three parent scales. By using our knowledge of parent scales, we will be able to bracket sections of the song together and focus on a single scale for the purpose of improvisation. We will not need to change modes or think about new chord tones quickly.

This tune has 32 bars with an AABC form. It is often played in either G minor or E minor; we will examine the G minor chords. The three parent scales are B♭ major, E♭ major, and G harmonic minor. We previously examined the chords and modes that can be derived from a major parent scale. Let's now examine the chords that can be derived from the harmonic minor parent scale. If we stack 7th chords on each note in the G harmonic minor scale we get the chords shown in Figure 2.61.

Figure 2.61

Chords derived from the G harmonic minor parent scale.

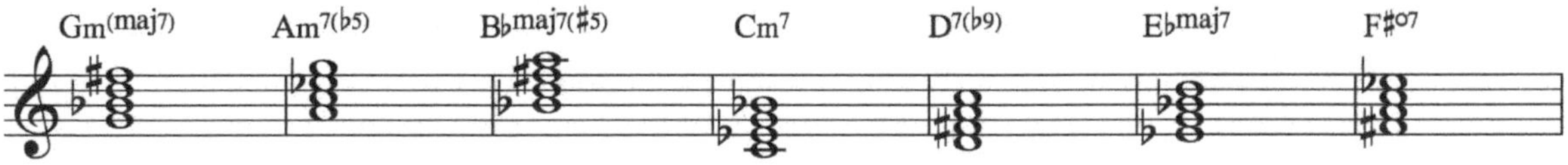

For this song we can use the chart below to organize the chords from each parent scale. Notice that both B♭ major and G harmonic minor both have an Amin$^{7\flat5}$ chord and a Cmin^{7} chord. We will use the context of each chord in the tune to figure out which parent scale would be most appropriate.

B♭ Major	**E♭ Major**	**G Harmonic Minor**
B♭Maj^{7}	E♭Maj^{7}	Gmin$^{(M7)}$
Cmin^{7}	Fmin^{7}	Amin$^{7\flat5}$ or Aø^{7}
Dmin$^{7(\flat9)}$	Gmin$^{7(\flat9)}$	B♭Maj$^{7\sharp5}$
E♭Maj$^{7(\sharp11)}$	A♭Maj$^{7(\sharp11)}$	Cmin^{7}
F^{7}	B♭^{7}	D$^{7\flat9}$
Gmin$^{7(\flat6)}$	Cmin$^{7(\flat6)}$	E♭Maj^{7}
Amin$^{7(\flat5)}$ or Aø^{7}	Dmin$^{7(\flat5)}$ or Dø^{7}	F♯dim^{7}

TIP

Reminder: The ii, iii, and vi chords will often be notated as a min7 chord without alterations so focus on the context of the chords to determine the appropriate parent scale.

The chords for the A section are shown in Figure 2.62.

Figure 2.62

Chords for the A section of "Autumn Leaves."

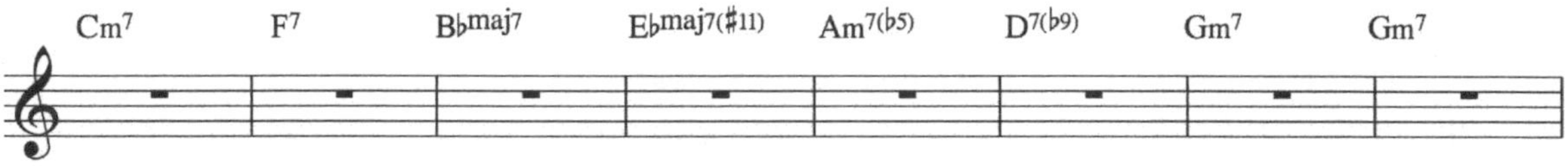

Look at the first four bars. All of these chords come from the B♭ major parent scale. So instead of thinking about four separate chords, we will bracket these bars with B♭ major and only play notes from that scale while we improvise.

Look at the last four bars. Although the B♭ major parent scale produces an Amin$^{7\flat 5}$ chord, we will consider Amin$^{7\flat 5}$ apart of the G harmonic minor parent scale because it leads to an altered dominant chord (D$^{7\flat 9}$ in this case) and then a minor tonic (Gmin^{7}). It is OK that the Gmin chord is a min^{7} and not a min$^{(M7)}$ chord. The main idea here is that the chord in the last 4 bars of the A section resolves to G minor and that G minor isn't a chord heading somewhere else. We can bracket bars 5–8 this time using only G harmonic minor.

TIP

Anytime you see a min$^{7\flat 5}$ chord, chances are you will use the harmonic minor parent scale, especially if it is followed by an altered dominant chord.

The chords for the B section are shown in Figure 2.63.

Figure 2.63

Chords for the B section of "Autumn Leaves."

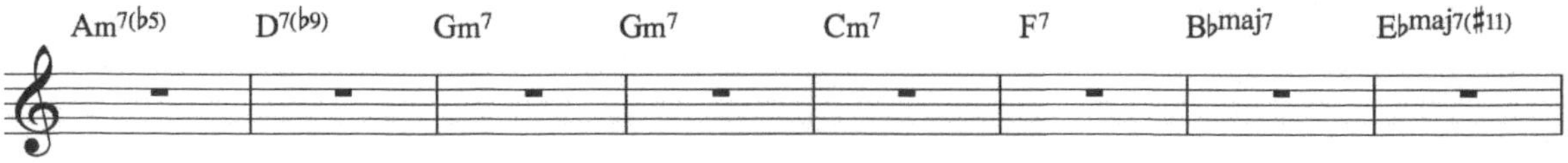

Do you notice that the B section uses the exact same chords as the A section but swaps the first 4 bars with the last 4 bars? Obviously, we can then bracket bars 1–4 with G harmonic minor and bars 5–8 with B♭ major.

The chords for the C section are shown in Figure 2.64.

Figure 2.64

Chords for the C section of "Autumn Leaves."

Bars 1–3 and 6–8 again use the ii, V, and i chords from G harmonic minor. We can essentially ignore the C^7 in bar 3 of the C section to simplify the progression since the main goal of bar 3 is to resolve to G minor. We can bracket these bars with G harmonic minor.

TIP

You could view the $Gmin^7$ to C^7 in bar 3 as part of the E♭Maj^7 progression. We often see a iii-vi-ii-V-I progression in music where the vi chord is either a min^7 or changed to a dom^7. In this case, I think the improvisation will be most clear for beginning improvisers if we focus on the resolution to G minor in bar 3.

Bars 4–5 use the ii, V, and I chords from E♭ major. We can bracket these bars with E♭ major.

Here is the entire tune with the bracketing concept (see Figure 2.65).

Figure 2.65

"Autumn Leaves" chord progression with the bracketing concept.

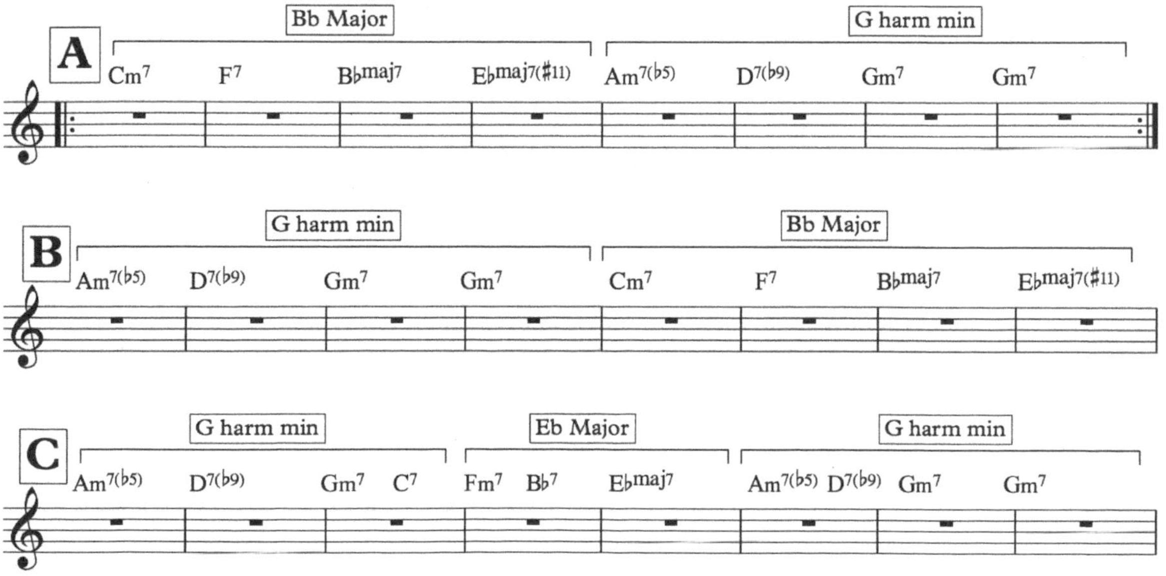

Creating an Accompaniment for "Autumn Leaves"

[▶] *See Example 2.4 on the companion website for a Band-in-a-Box file and audio recording example of the "Autumn Leaves" activity.*

Let's create an accompaniment in Band-in-a-Box for "Autumn Leaves." This tune uses fairly complex chords so we will have to be careful of how we enter a min$^{7\flat5}$, dom$^{7\flat9}$, and Maj$^{7\#11}$. I will discuss how to enter those chords with the QWERTY keyboard and a special feature called the Chord Builder.

1. Launch Band-in-a-Box.
2. Click in the *Title* window. It will display "Untitled" when you open a new Band-in-a-Box file.
3. In the *Main Settings* dialogue box that pops up, enter the title (Autumn Leaves), key (Gm), and a moderately slow tempo (120 bpm). It is fine to keep *Embellish Chords* checked for this song.
4. Click OK.
5. The length of the song should already be set to 32 since this is the default value. If it is not set to 32, click on the box that says "End" and select 32.
6. Set the number of choruses to 5. This will allow us to perform the melody during choruses 1 and 5 and improvise during choruses 2–4.
7. Click the actual number at bar 1 twice to turn off the feel change. The number shouldn't be blue or green anymore.
8. Click on the "C" in bar 1. Type "Cm7" on your QWERTY keyboard.
9. Click on the empty space in bar 2. Type "F7" on your QWERTY keyboard.
10. Click on the empty space in bar 3. Type "BbJ" on your QWERTY keyboard.

TIP

"J" is a shortcut in Band-in-a-Box for a Maj7 chord.

11. Click on the empty space in bar 4. Select *User > Chord Builder*. In the Chord Builder dialogue box, select "E♭" as the root. Then click the pop-up next to *Other* and select Maj$^{9\#11}$ as the chord quality. Click *Enter Chord*. See Figure 2.66.
    a. Alternatively, you can also type "EbMaj9#11" on your QWERTY keyboard to get the same chord. Band-in-a-Box does not recognize the Maj$^{7\#11}$ quality.

Figure 2.66
Chord Builder window for E♭Maj$^{9\sharp 11}$.

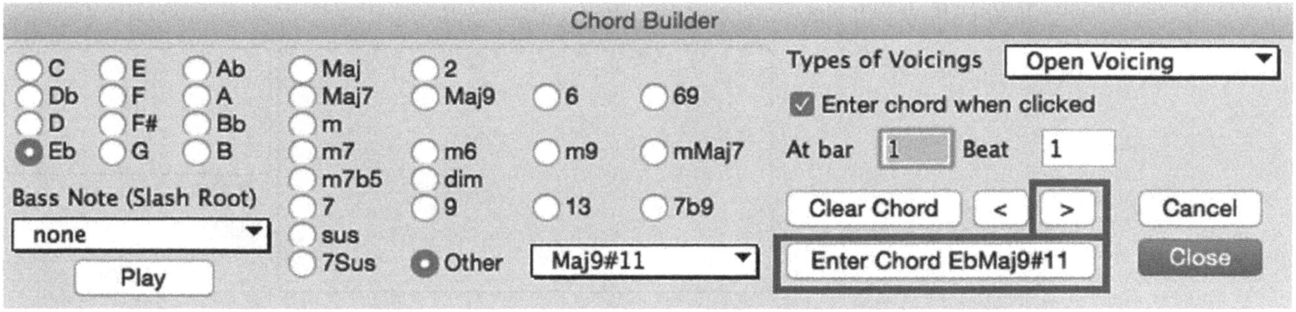

TIP

A chord instrument player will often interpret a Maj7♯11 chord to include the 9th so Maj7♯11 and Maj9♯11 are essentially the same chord.

12. Still using the Chord Builder, click the right facing arrow to advance to the start of bar 5. Select "A" as the root and "m7b5" as the quality (see Figure 2.67). Click *Enter Chord.*

Figure 2.67
Chord Builder window for Am$^{7\flat 5}$.

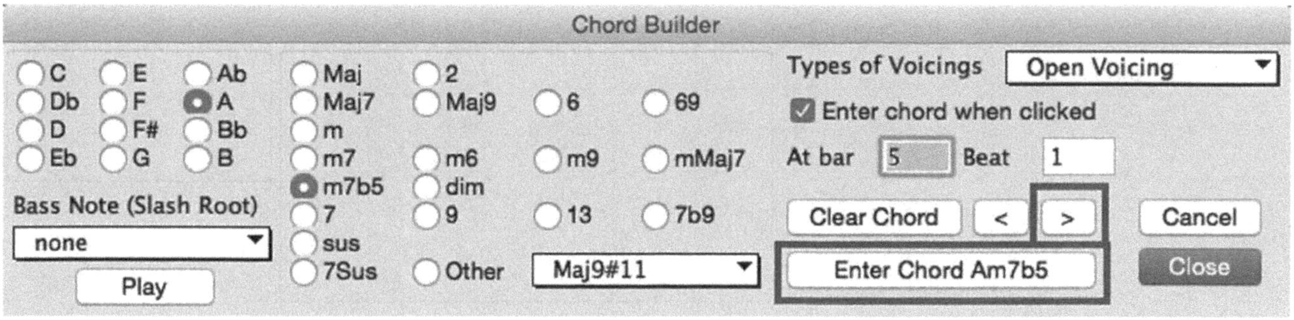

13. Still using the Chord Builder, click the right facing arrow to advance to the start of bar 6. Select "D" as the root and "7♭9" as the quality (see Figure 2.68). Click *Enter Chord.*

Figure 2.68

Chord Builder window for $D^{7\flat9}$.

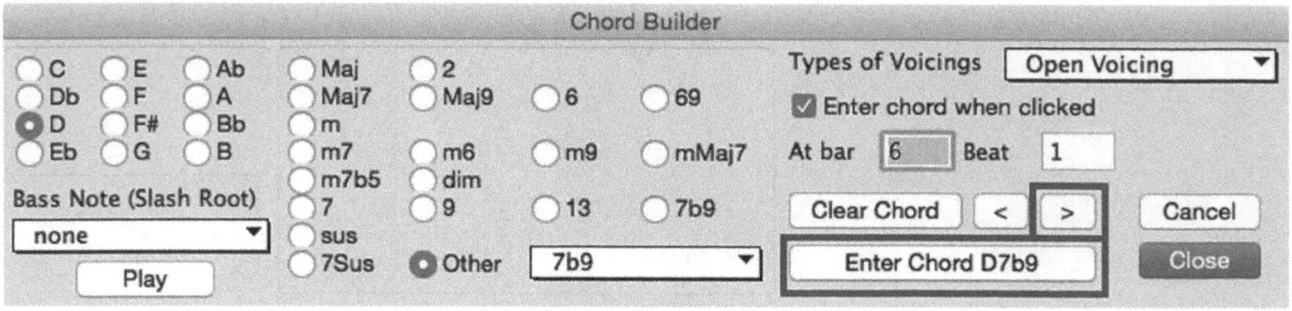

14. Click *Close* to exit the Chord Builder.
15. Click the empty area next to bar 7 and type "Gm7" on your QWERTY keyboard.
16. Click and drag to select bars 1–8. Select Edit > Copy.
17. Click the empty area next to bar 9. Select *Edit > Paste*. You should now have both A sections completed (see Figure 2.69).

Figure 2.69

"Autumn Leaves," bars 1–9 entered into Band-in-a-Box.

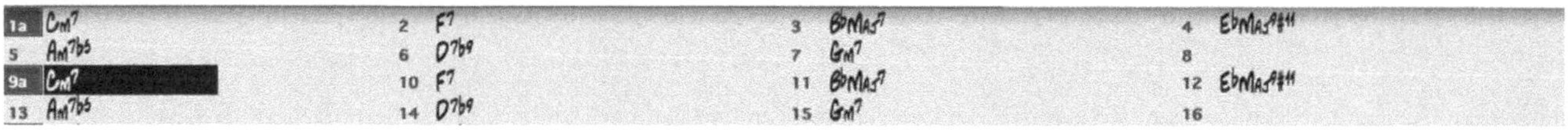

18. Click and drag to select bars 5–8. Select *Edit > Copy.*
19. Click the empty area next to bar 17. Select *Edit > Paste.*
20. Click the empty area next to bar 25. Select *Edit > Paste.*
21. Click and drag to select bars 1–4. Select *Edit > Copy.*
22. Click the empty area next to bar 21. Select *Edit > Paste.*
23. Click the second half of bar 27 and type "C7" on your QWERTY keyboard.
24. Click the empty area next to bar 28 and type "Fm7" on your QWERTY keyboard.
25. Click the second half of bar 28 and type "Bb7" on your QWERTY keyboard.
26. Click the empty area next to bar 29 and type "EbJ" on your QWERTY keyboard.
27. Copy/paste a $\text{Amin}^{7\flat5}$ and a $D^{7\flat9}$ chord into bar 30.
28. Copy/paste a Gm^{7} chord into bar 31.
29. Set the A sections to the "A" feel of the the style by clicking the bar number at bar 1 and 9.
30. We will use the "B" feel of the style in the B and C sections. Click the bar number twice at bar 17 and 25. See Figure 2.70 for a screenshot of the chord progression entered into Band-in-a-Box.

Figure 2.70

Entire "Autumn Leaves" chord progression entered into Band-in-a-Box.

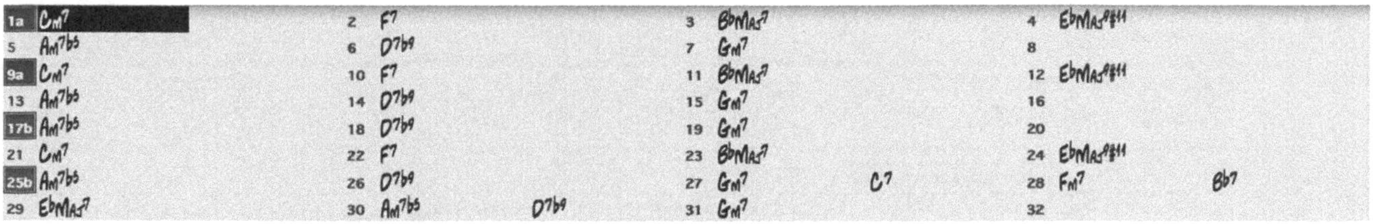

31. Play the accompaniment. You should hear a feel change at bar 17, a drum fill into bar 25, and a continuation of the "B" feel to the end.
32. Try selecting various styles by clicking the *Style* icon.
33. If you have a MIDI keyboard connected to your computer, record in the melody. Listen to and evaluate your performance. Re-record if necessary. Once you have a good take, check the box for *Copy 1st Chorus to Whole Song* and click *OK - Keep Take.*
34. Now we're ready to practice our improvisation. The goal is to perform the appropriate parent scale throughout each bracketed section. Confine yourself to only playing notes from the parent scale. It is a good idea to slow down the tempo to get comfortable with the scales.
    a. Practice the parent scales.
        i. Practice B♭ major, G harmonic minor, and E♭ major on their own. I suggest running eighth notes up and down the scale once (see Figure 2.71).

Figure 2.71

B♭ major, G harmonic minor, and E♭ major scales in eighth notes from the root.

        ii. Practice starting each scale from different scale degrees. It is good to avoid always starting the scale from the parent root. For example, play the B♭ major scale starting on C followed by the G harmonic minor scale starting on C and the E♭ major scale starting on C (see Figure 2.72).

Figure 2.72

B♭ major, G harmonic minor, and E♭ major scales in eighth notes from a common note.

b. Practice playing the modes using eighth notes up and down the scale along with the accompaniment at a slow tempo. Slowly speed up the tempo with a goal of approximately 120 bpm.

c. Practice playing various permutations of the mode such as the scale in 3rds along with the accompaniment.

d. As you play along with the accompaniment be sure to keep your ears wide open. Although a B♭ major scale works over the first four measures, each note of the scale will feel different over each chord. Let your ear guide the tension notes to resolution. Very often a resolution note on one chord will be a tension note on the next chord. For example, I could play a B♭ note over the Cmin7. The B♭ has a feel of resolution on the Cmin7 chord. If I hold that B♭ into the F7 chord, it will be a tension note and I can resolve it down to the A which is the 3rd of the F7 chord (see Figure 2.73). The more you experiment with the parent scales on this progression, the better your ear will get at guiding you to the best sounding notes for each chord.

Figure 2.73

Resolving a tension note (B♭) to a release note (A) over F7.

35. Don't forget to focus on phrasing. Leave space for the listener to digest your musical ideas.

a. When starting with "Autumn Leaves," I believe the best phrasing structure is to play a 3-bar phrase followed by a 1-bar rest (see Figure 2.74). This will

Figure 2.74

"Autumn Leaves" phrasing structure; 3-bar phrase followed by a 1-bar rest.

emphasize the 4-bar phrase structure of the tune and allow your brain to reset for each parent scale.

36. Band-in-a-Box includes a feature called the Soloist Generator. This very cool feature will actually create an improvised solo based on the style of various famous players or within a particular style. For example, there is a soloist file called "Bebop Saxophone" and another called "J Henderson" (named for Joe Henderson, the great jazz tenor saxophonists). Open the Soloist Generator by selecting *Soloist* > *Generate and Play a Solo*. See Figure 2.75 for a screenshot of the Soloist Generator window.

Figure 2.75

Soloist Generator window in Band-in-a-Box.

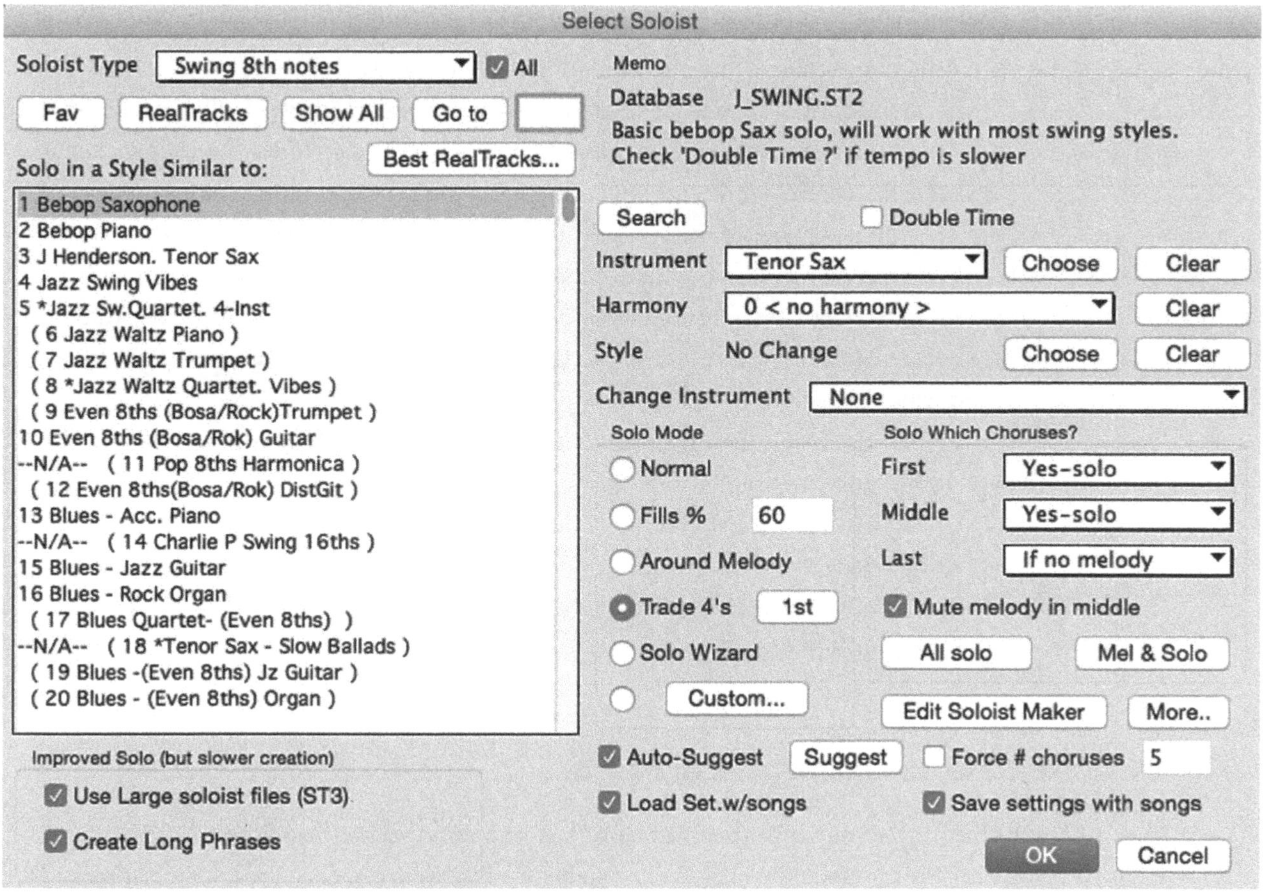

37. Click on a soloist in the left pane of the Soloist Generator dialogue box. Click OK to generate the solo.
38. Listen and evaluate the soloist. Try out other soloists until your find a soloist file that you really like.
39. View the notation of your solo.
    a. Click the *Notation* icon. 
40. Click the "S" for soloist. 

41. Open the Soloist Generator once more. Explore the fine-tune adjustments that can be made to the soloist such as selecting a new instrument, changing the instrument throughout the song, selecting the soloist choruses, and more.

TIP

I especially like selecting *Edit Soloist Maker* in the Solo Generator window and exploring the new dialogue box. In this window you can restrict the range of the soloist. This is especially helpful for younger brass students who lack range. The range is given in MIDI note numbers. Middle C is MIDI note number 60. As you go up 1 semitone you add 1 number so C♯ above middle C is MIDI note number 61.

42. When improvising it is often fun to "trade" phrases with another soloist. The most common way to trade is to trade 4's, meaning one player will improvise for 4 bars and then the next player will solo for 4 bars and so forth. This allows the soloists to build a musical conversation and provide inspiration for new ideas. The Soloist Generator can create a trading situation where you can perform the first 4-bar phrase and the Band-in-a-Box soloist will play the next 4-bar phrase. In the Soloist Generator dialogue, click the button next to *Trade 4's.* Click the button labeled *1st* until it says *2nd* to tell the Band-in-a-Box soloist to play during the second phrase. Click OK. 

43. Practice improvising on your instrument along with Band-in-a-Box trading 4's throughout the song.
44. Go back to the Soloist Generator dialogue box and let Band-in-a-Box play the first phrase so you have a chance to practice the parent scales that occur during the last 4 bars of each section. 

45. Save your Band-in-a-Box file to a memorable location.

iPad Connection: iReal Pro

Links: https://itunes.apple.com/us/app/ireal-pro-music-book-play/id298206806?mt=8.

iReal Pro by Technimo LLC is available for $12.99 from the Apple App Store for iOS devices. I will focus on the iPad version, but the app works equally well on an iPhone or iPod touch. Although it lacks some of the advanced features of Band-in-a-Box such as the ability to generate a solo, at only $12.99 it is a great way to create accompaniments on the iPad at a relatively low-cost.

Although we started in Band-in-a-Box by entering chords to create custom accompaniments from scratch, I can often find pre-made user-created accompaniments for most tunes using the iReal Pro Forum. The Forum is a section of the app that connects a huge community of users together both for downloading accompaniments and posting questions. I suggest becoming a registered user so you can submit questions and share completed accompaniments to the forum.

[▶] *See Example 2.5 on the companion website for an iReal Pro file, PDF, and audio recording example of the "iReal Pro" activity.*

Downloading a Tune from the iReal Pro Forum

1. Hold the iPad in the vertical orientation and tap iReal Pro on your iPad to launch the app.
2. Tap *Forums* in the lower left corner. If you don't see the icon, tap the center of the screen to open the side pane.
3. Swipe up to navigate to the bottom of this window and tap *Register*. Create a username and password. After going through the registration process, you will need to check your email and click on the link to activate your registration. If you have already registered previously, you can tap on *Log In* and enter your credentials.

TIP

You do not need to register to search and download accompaniments. Registration allows you to post questions and your own accompaniments to the forum.

4. Tap the *Search* icon to start a new search. Enter a tune name such as "Autumn Leaves" and tap the blue magnifying glass icon.
5. You will most likely receive many results. Most of the results will be in a group of other tunes. For example, when I searched "Autumn Leaves" one of the results was titled "Jazz Standard Bible."
   a. To explore the items included in a result, tap the arrow. This will take you to a detailed view of the result. Most results will include a full bundle download as well as individual tune downloads.
   b. Tap either the full bundle or an individual song to download and install the tunes.

TIP

The forum has a wide variety of accompaniments from every genre. Try searching for pop artists (Michael Jackson, The Beatles, etc.) or great classical composers (Bach, Beethoven, Mozart, etc.). I found some interesting classical accompaniments while searching "Bach" in the forum. NOTE: iReal Pro currently only has accompaniment styles in the jazz, latin, and pop genres so the classical accompaniments will be played in a jazz/latin/pop feel. Band-in-a-Box does have classical styles. Hopefully iReal Pro will continue to add to their style options and eventually include additional genres such as classical.

6. Tap the *Songs* icon in the lower left corner. Browse through the songs in your library and confirm that the song you downloaded in the previous step resides in your library.

TIP

Tap the *Search* box at the top to search through your library.

TIP

If you downloaded a full bundle, you can tap the *Playlists* icon in the lower left corner to view the bundled playlist. For example, I downloaded the bundle called "Beatles" and this bundle now appears in my playlists with 37 Beatles accompaniments (see Figure 2.76). It is faster for me to find these tunes from the Playlists menu than the songs menu.

Figure 2.76
37 Beatles tunes in a *Beatles* playlist downloaded from the iReal Pro Forum.

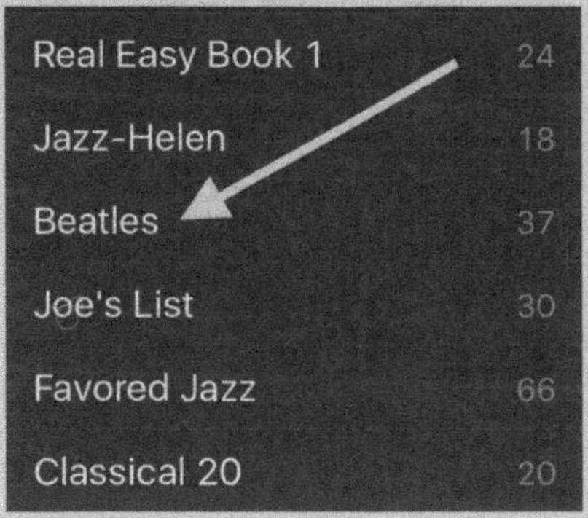

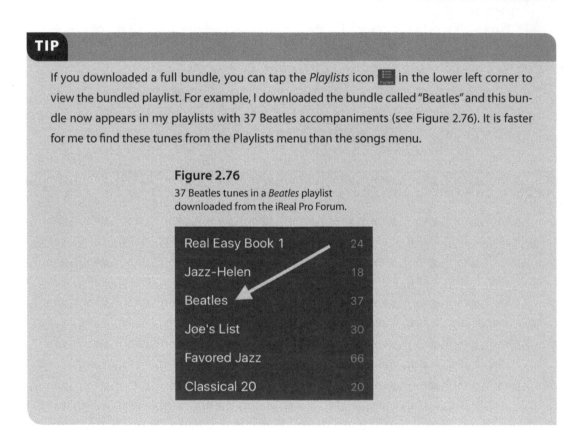

7. Tap the song you downloaded to open it.
8. Tap the play button in the bottom center of the screen.

Modifying an Existing Accompaniment

1. To select a new style, tap the style name at the bottom left corner of the screen (ex. Jazz - Medium Swing. This opens up a window with styles in the jazz, latin, and pop genres. The app designers have done a great job of adding new styles to the app. I particularly enjoy "Blue Note" under "Jazz" to get

that grooving feel on tunes like Herbie Hancock's "Cantaloupe Island" or "Watermelon Man."

2. Adjust the tempo by tapping the tempo icon 100 in the bottom center of the screen. You can adjust the tempo by dragging the tempo slider or tapping the "+" and "–" icons. You can also generate a "tap tempo" by tapping the metronome icon in the desired tempo.
3. Adjust the number of choruses or repeats by pressing the *Repeats* icon 3x. Similar to the way you use the tempo window, you can adjust the number of repeats by dragging the slider or tapping the "+" and "–" icons.
4. Transpose the song to the desired key by pressing the *Transposition* icon G-. The keys go up and down chromatically.

TIP

If you are playing on a transposing instrument tap the *Gear* icon in the top right corner and then tap *Transposing Instrument* in the pop-up menu. For example, tenor saxophone is a B♭ instrument so I could tap "B♭" to visually transpose the chords up a major 2nd interval. NOTE: This is only for *visual* transposition.

5. Tap the Chord Diagrams icon. This is my second favorite feature in iReal Pro next to the Forum! iReal Pro can actually display information about how to play chords and which scale is appropriate for the given chord as it plays the accompaniment in real time. You can choose from guitar chords, piano chords, ukulele chords, and chord scales. Note: After you select the appropriate chord diagram, you must play the accompaniment to have the diagrams display on-screen.
   a. Guitar Chords: Display left- or right-handed diagrams. Explore the library (tap the *Library* button) to see guitar chord diagrams in every key and quality.
   b. Piano Chords: Display one- or two-handed voicings. Explore the library (tap the *Library* button) to see a variety of chord voicings in every key and quality. For example, iReal displays 4 different voicings for CMaj7.
   c. Ukulele Chords: Similar to guitar chords but for ukulele.
   d. Chord Scale: Display the appropriate scale/mode for each chord in treble or bass clef. Wouldn't it be nice to have the Cmin^{7} chord and C dorian scale displaying on the screen while you try to improvise (see Figure 2.77)?

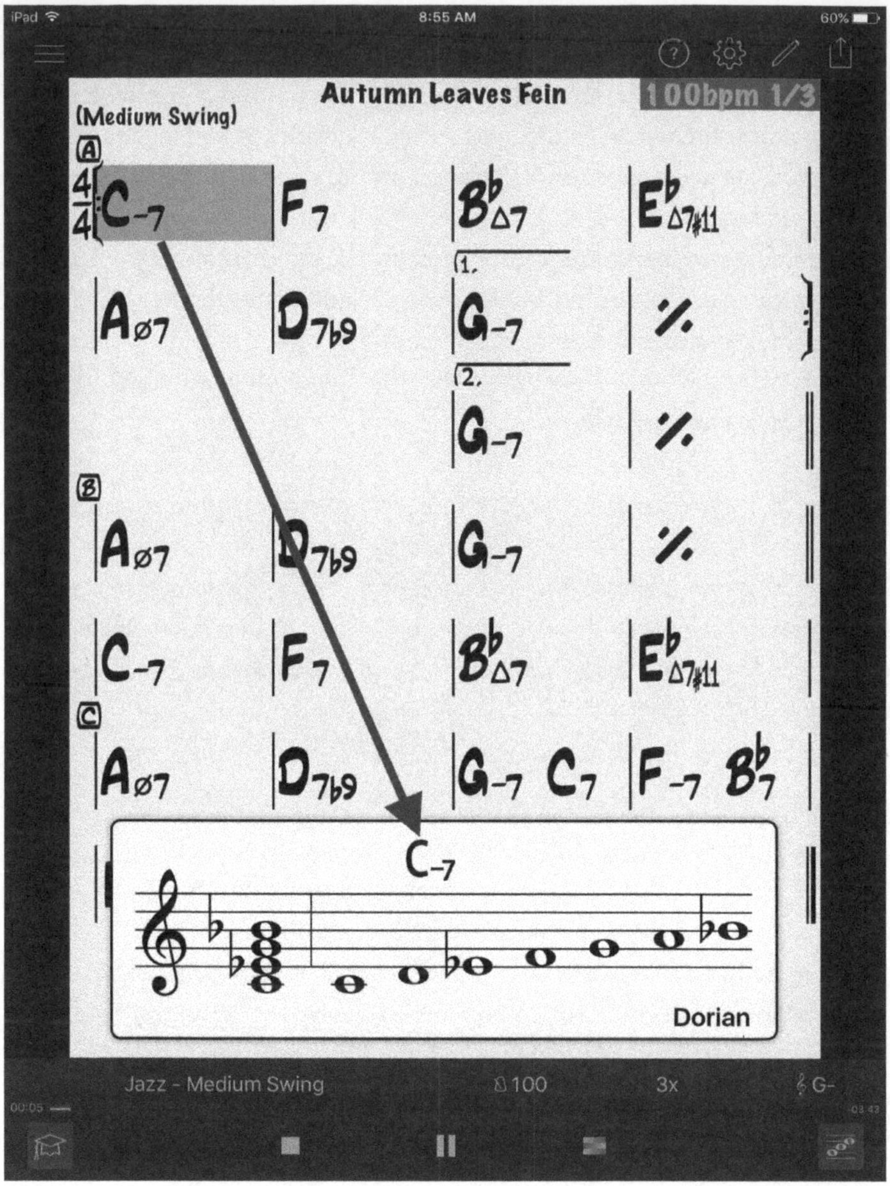

Figure 2.77 iReal Pro displays the chord/scale for Cmin⁷.

6. Tap the Mixer icon. Adjust the volume of each instrument (chords, bass, drums, and reverb) by dragging the volume slider for each instrument. Mute an instrument by pressing the "x" icon on the left side of each instrument. Select a different instrument by tapping the treble clef, bass clef, or drum icon. Each style has a different set of instruments from which you can select. For example, in the *Jazz - Medium Swing* style, the chordal instrument can be piano, rhodes 1, rhodes 2, vibraphone, or organ.
7. Tap the *Practice* icon. As the accompaniment plays, iReal will automatically modify the playback as desired during each chorus. For example, if I set the tempo to +5 bpm, iReal will increase the tempo by 5 bpm each chorus. If I set the repeats to 10, I could start practicing the tune at 80 bpm and end up at 125 bpm without having to start or stop the accompaniment. The transposition practice feature is great for practicing tunes or chord progression in all 12 keys.

8. Opening an existing accompaniment is usually very helpful, but the quality of the accompaniment always depends on the user who entered the chords. If you want/need to make changes to the chord progression, tap the *Pencil* icon in the upper right corner.
   a. Tap *Edit.*
   b. Tap an existing chord. The cursor will flash and you will see the chord indicated in the text window. Modify the chord as needed. For example, to change an E♭Maj$^{7\sharp 11}$ to E♭^{7} just delete everything except the E♭ and add a 7.

TIP

If you enter a chord quality and it shows up blue in the big window, you have entered a chord that iReal Pro can't understand. Recheck your entry or modify your chord. Do you recall when Band-in-a-Box didn't understand Maj$^{7\sharp 4}$?

TIP

iReal uses "Δ" for major, "-" for minor, "ø" for min$^{7\flat 5}$, "o" for diminished, and "+" for augmented.

   c. Tap an empty area to add an extra chord. For example, I open a tune with a Cmin^{7} for a bar but I would prefer to have the Cmin^{7} for 2 beats and F^{7} for 2 beats. I would tap in the middle of the bar and enter an F^{7} chord.
   d. Tap *Save* in the upper left corner and *Done* in the upper right corner after you have made the necessary changes.

Entering "Autumn Leaves" from Scratch

In the rare instance that you can't find a given song in the Forum or if you just want to create an original tune accompaniment, you can generate an accompaniment from scratch. I will walk you through entering "Autumn Leaves" into iReal Pro.

1. Tap *Songs* in the upper left corner to open the side pane.
2. Tap the *Songs* icon in the lower left corner.
3. Tap the "+" icon in the top area of the songs pane.
4. Choose from one of the templates or, if your song doesn't fit the given templates, just select *Blank*. In the case of "Autumn Leaves," I suggest selecting the 32-bar AABA form even though the form of the song is really AABC. It is easier to start with the template and make minor changes.
5. Tap the *Info* icon in the center right of the screen. Enter the tune name (*Autumn Leaves*). Select a style. Set the key (G minor).
6. Tap C - 7.

TIP

You can tap the *Menu* icon to the left of to the chord text input to open the chord quality menu. This menu will only be available after you enter the root note.

7. Tap *Next* four times to advance to the next measure.
8. Tap F 7.
9. Tap *Next* four times to advance to the next measure.
10. Tap B b Δ 7.
11. Tap *Next* four times to advance to the next measure.
12. Tap E b Δ 7 # 11
13. Tap *Next* four times to advance to the next measure.
14. Tap A ø 7.
15. Tap *Next* four times to advance to the next measure.
16. Tap D 7 b 9.
17. Tap *Next* four times to advance to the next measure. You should now be under the 1st ending.
18. Tap G - 7.
19. Tap *Next* four times to advance to the next measure.
20. Tap the measure repeat symbol.
21. Tap at the start of the measure in the 2nd ending.
22. Tap G - 7.
23. Tap *Next* four times to advance to the next measure.
24. Tap the measure repeat symbol.

Your "A" section should look like Figure 2.78.

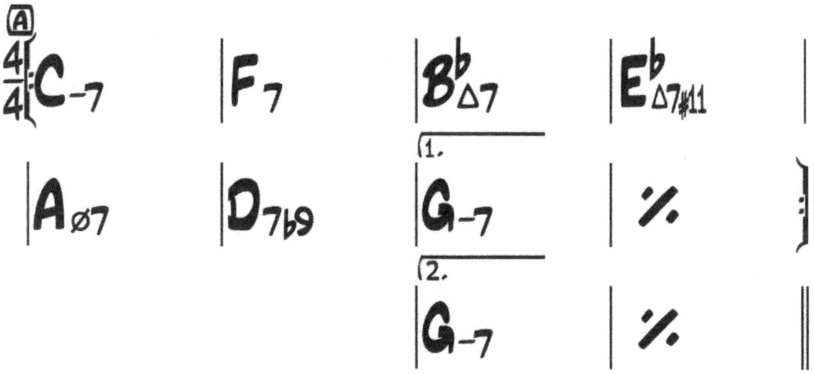

Figure 2.78
"Autumn Leaves""A" section entered into iReal Pro.

25. Double-tap and drag across the last 4 bars of the A section. As you drag you should see a blue highlight over the selected area.
26. Select *Copy* (see Figure 2.79).

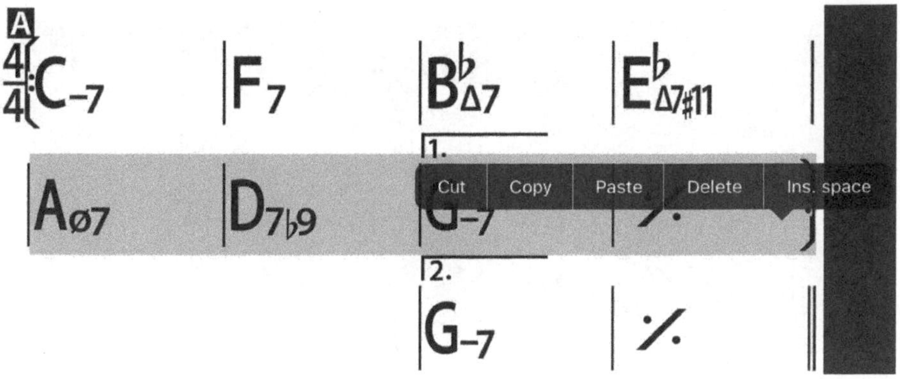

Figure 2.79
Double-tap and drag across the last 4 bars of the A section and select *Copy*.

27. Double-tap in bar 11 (the first bar of the B section) and drag across to the end of bar 14.
28. Select *Paste.*
29. iReal Pro copies EVERYTHING including the 1st ending, repeat, and bar lines.
    a. Tap bar 11, beat 1. In the chord entry area, select *Symbols.* It is located in the area of the spacebar on the QWERTY keyboard.
    b. Tap the left single barline icon and the Rehearsal B icon (see Figure 2.80).
    c. Tap bar 13, beat 1. In the chord entry area, tap the 1st ending icon four times until the symbol disappears (see Figure 2.80).
    d. Tap bar 14, beat 4. In the chord entry area, tap the right single barline icon (see Figure 2.80).

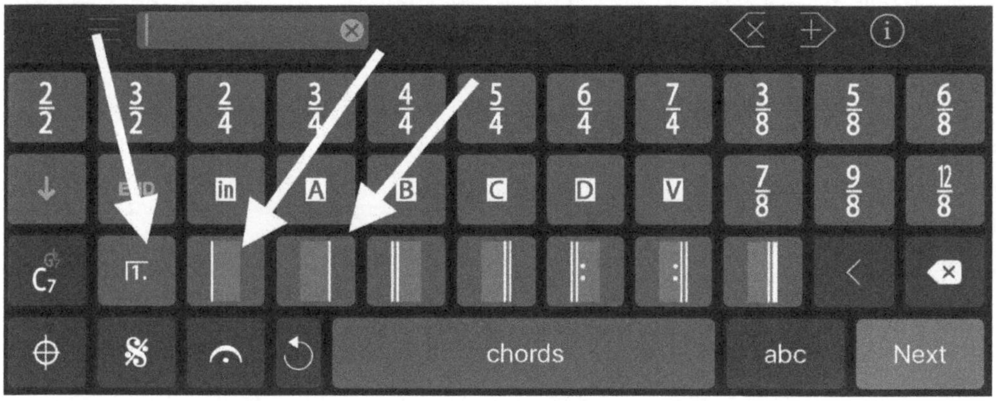

Figure 2.80
1st ending, left barline, and right barline in the *Symbols* menu.

30. Double-tap and drag across the first 4 bars of the A section.
31. Select *Copy*.
32. Double-tap in bar 15 and drag across to the end of bar 18.
33. Select Paste.
34. Again, we have some cleanup to do.
    a. Tap bar 15, beat 1. In the chord entry area, tap the 4/4 icon, Rehearsal A icon, and left single barline icon (see Figure 2.81).
    b. Tap bar 18 beat 4. In the chord entry area, tap the right double barline icon (see Figure 2.81).

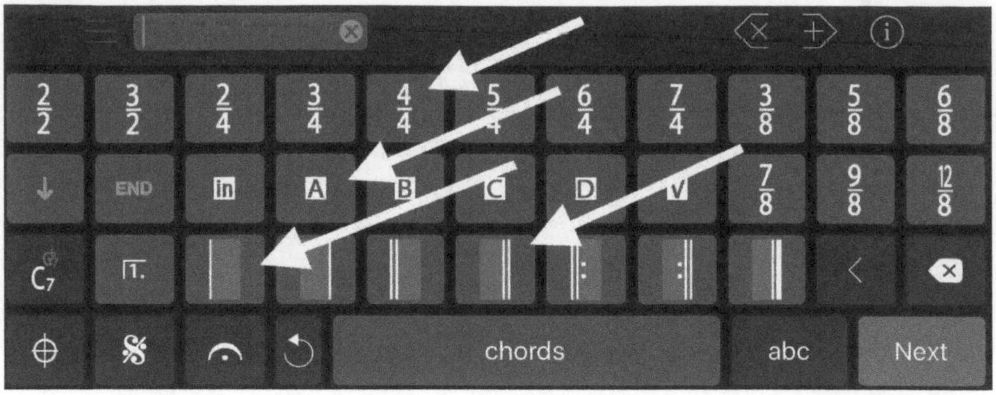

Figure 2.81
Rehearsal letter, time signature, left barline, and right double barline in the *Symbols* menu.

35. Double-tap and drag to select bar 11 to bar 13 beat 2.
36. Select Copy.
37. Double-tap in bar 19 and drag across to bar 21 beat 2.
38. Select *Paste.*
39. Tap bar 19 beat 1. In the chord entry area, tap the Rehearsal C icon.
40. Tap bar 21 beat 3.
41. Tap the *Chords* icon (same place as the Symbols icon where Spacebar would be) to return to the chord entry palette.
42. Tap C 7.
43. Tap *Next* twice to advance to bar 22.
44. Tap F - 7.
45. Tap *Next* twice to advance to bar 22 beat 3.
46. Tap B b 7.
47. Tap *Next* twice to advance to bar 23
48. Tap E b Δ 7.
49. Tap *Next* four times to advance to bar 24.
50. Tap A ø 7.
51. Tap *Next* twice to the advance to bar 24 beat 1.
52. Tap D 7 b 9.
53. Tap *Next* twice to advance to bar 25.
54. Tap G - 7.
55. Tap *Next* twice to advance to bar 26.
56. Tap the measure repeat icon.
57. Tap *Done* in the upper right corner and select *Save.*

Your complete chord progression should look like Figure 2.82.

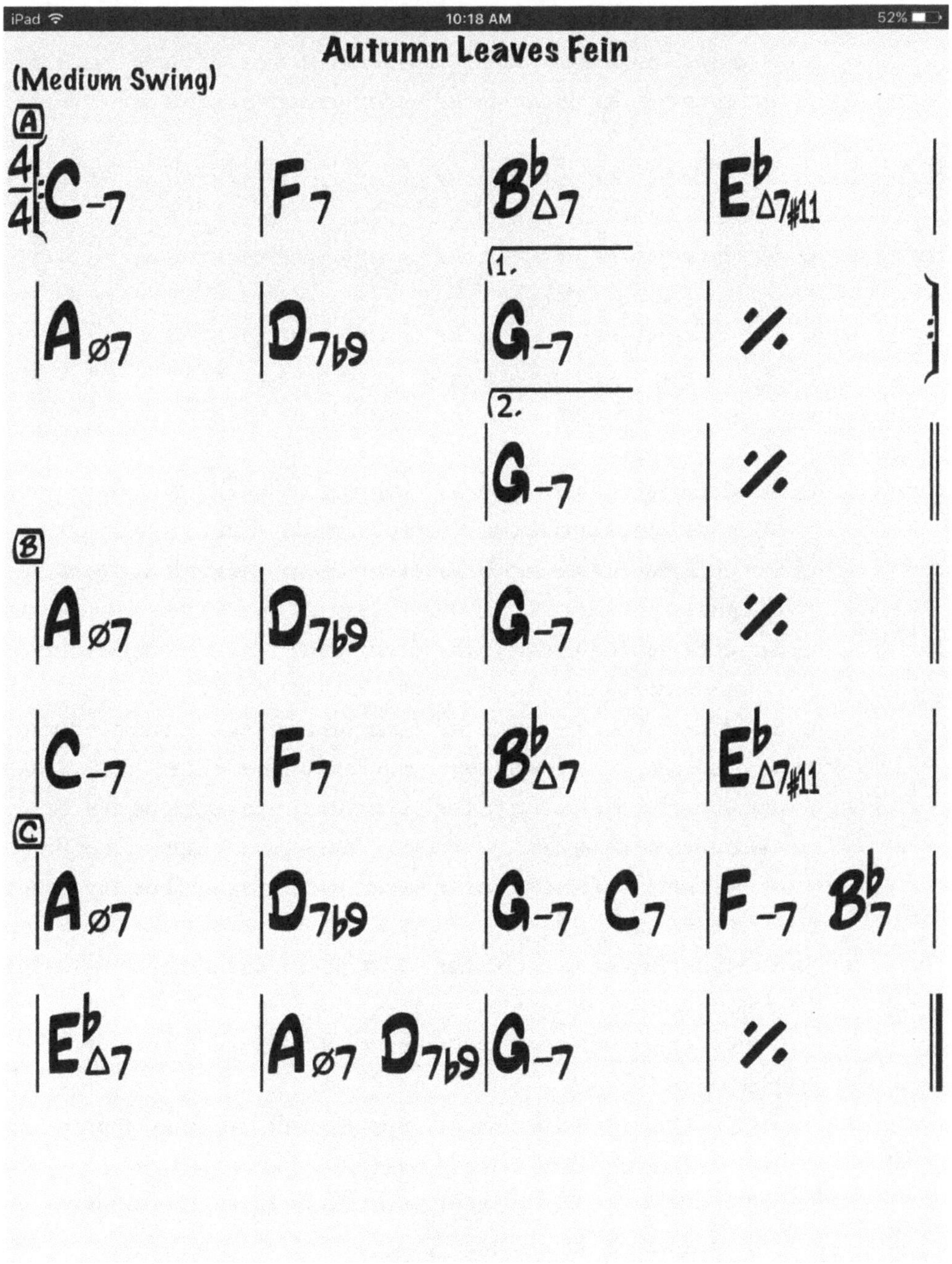

Figure 2.82
Entire "Autumn Leaves" chord progression entered into iReal Pro.

58. Tap the *Share* icon in the upper right corner.
    a. Share Chord Chart
        i. Select *iReal Pro Format* for importing the accompaniment file into iReal Pro on another iOS device. You can mail or save the html file to Dropbox or Google

Drive (I will discuss Google Drive in Ch. 6). Open the email, Dropbox link, or Google Drive link on your iOS device and tap the highlighted song in the window to open and import the accompaniment into iReal Pro.

TIP

To open an iReal Pro Format chord chart using Google Drive:

- Locate the file in your Google Drive app.
- Tap the file and select *Open In*.
- Tap *Open In* again in the next window and then select *Copy to iReal Pro*.

ii. Select *PDF* to create an image file of the lead sheet. This can then be printed from a computer or added to other apps on your iPad such as Dropbox, Google Drive, or forScore. The forScore sheet music app is discussed in Chapter 3.

iii. Select *Forums* to share your accompaniment with the entire iReal Pro community. Sharing is good!

iv. Select *Music XML* to open the accompaniment in music notation software such as Finale or Sibelius on your computer. When you open this XML file in notation software, you'll get a single staff with chords above the staff. This is mostly helpful if you need to make some minor layout changes for printing. For example, if you wanted to have 8 bars per system this could be easily achieved in notation software.

b. Share Audio: Export as audio when you want to send the audio playback of the accompaniment to a student who does not have the iReal Pro app.

i. WAV—Large audio files (approximately 10mb/min) appropriate for burning to CD.

ii. AAC—Compressed audio files (approximately 1mb/min) appropriate for emailing or posting to the internet.

iii. MIDI—Very tiny files appropriate for importing into other music software such as GarageBand, Finale, or Sibelius.

TIP

iReal Pro has done a great job of incorporating help directly in the app. Tap the *Help* icon in the top right corner of the screen.

- Tap *Tutorials* to access excellent short tutorials.
- Tap *Help & Support* to explore FAQs, search the help forum, or send an email for assistance.

Chapter 2 Review

1. What scales work for the entire blues progression? What key allows the player to use only black keys while improvising on the basic blues progression? Can you

think of any other examples in music when dissonance between scale/melody and chord are acceptable?

2. This chapter discussed the concept of phrasing and space in an improvised solo. Select a specific recording that includes improvisation and discuss the phrasing concepts employed by the artist. For example, discuss the phrasing used throughout the first chorus of Miles Davis's improvised solo on "So What" from the album *Kind of Blue*.
3. Band-in-a-Box includes the Wizard and the Soloist Generator to help with improvisation. Describe the pros and cons of these two features.
4. Band-in-a-Box includes many styles and soloists. List your favorites styles and soloists included in Band-in-a-Box and describe why they are you favorites.
5. Examine the chord progression for "Autumn Leaves" again. Instead of bracketing the first four measures of the A section as "B♭ major parent scale," identify the name of the specific mode for each chord
   a. $C\text{-}^{7}$: ____________________
   b. F^{7}: ____________________
   c. $B\flat Maj^{7}$: ____________________
   d. $E\flat Maj^{7\sharp11}$: ____________________
6. What is the parent scale for $Amin^{7\flat5}$ in the following examples and why?
   a. $Amin^{7\flat5}$ - $D^{7\flat9}$ - Gmin
      i. Parent Scale: ____________________
      ii. Why? ____
      ____
      ____
   b. $Amin^{7\flat5}$ - $Dmin^{7}$ - $Gmin^{7}$ - $Cmin^{7}$ - F^{7} - $B\flat Maj^{7}$
      i. Parent Scale: ____________________
      ii. Why? ____
      ____
      ____
7. Create a venn diagram comparing Band-in-a-Box and iReal Pro as tools for teaching improvisation (see Figure 2.83).

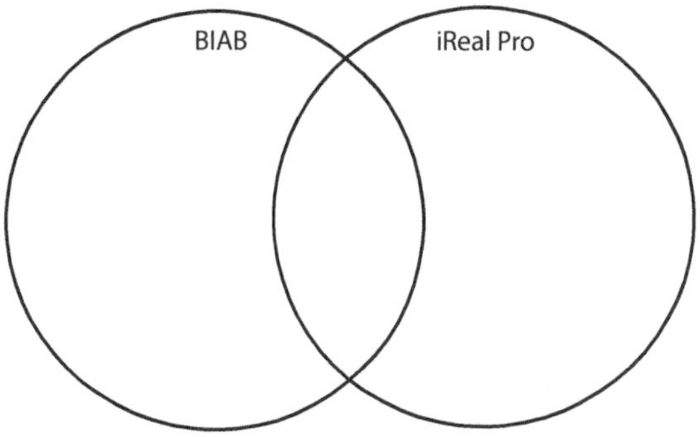

Figure 2.83
Venn diagram for comparing Band-in-a-Box and iReal Pro as tools for teaching improvisation.

3 NOTATION SOFTWARE

Note Flight and forScore

I often hear teachers less comfortable with technology throw their hands up and say, "I can handwrite music faster than I can do it on the computer!" True, music notation applications do have a learning curve; however, we must remember the many advantages notation software affords us. By creating digital music notation, teachers can quickly transpose music (both notes and chords!) for lead sheets and exercises sheets, copy and paste to save time entering repetitive melodies/exercises, import MIDI from other software (such as Band-in-a-Box), and share exported PDFs with students. This chapter will also discuss how to organize music on the iPad with forScore. As a gigging musician, I no longer carry a crate of Fake Books or a 3" binder full of transcribed charts in plastic sleeves; instead, I create all digital versions of my music and store them on an iPad for easy sorting, access, and sharing—not to mention, I no longer need a stand light!

Notation Software Options

Finale and Sibelius are the two biggest players in the world of notation. Both applications are excellent and you will find staunch supporters of each in the music tech world. Cost is the primary factor that limits the availability of the full version of Finale and Sibelius to students. Starting at approximately $250 and up for new users, "big boy" notation is expensive. Fortunately, both companies offer introductory versions of their software with slightly limited features. These titles are Finale NotePad (free) and Sibelius First ($99) and both will generally provide the features needed for all the activities explored in this text.

The other newer application available is Note Flight and this company operates on a completely different business model. Note Flight is a FREE application that runs on Mac, Windows, iPad, Chromebook, etc. (basically anything with a web browser!). Students do not need to download or install anything to work with Note Flight; they simply need to be able to open a web browser and access www.noteflight.com. Once at noteflight.com, students can create an account with an email address. For additional features, Note Flight offers a premium membership that costs $49/year for an individual or classroom packages (called Note Flight Learn) starting at $69/year for 10 users (plus $2 more for each additional user, so access for 20 total users costs $89, for 30 total users it costs $109, etc.).

TIP

Sometimes, especially in elementary school, it is not possible to have each student login using a personal email address. Instead, teachers can create multiple accounts and assign students to a particular account for the year. For example, I may create *Haverford_student01, Haverford_student02,* etc. using multiple Gmail addresses that I create for this purpose. Of course, the other option is to use Note Flight Learn (paid subscription) to manage your classes/students.

For the purposes of this text, I will provide specific step-by-step procedures for Note Flight since the interface is identical on all devices (Mac/Windows/iPad/Android tablet/Chromebook/etc.).

Creating a Lead Sheet: "Für Elise" by Beethoven

[▶] *See Example 3.1 on the companion website for XML and PDF example files for the "Für Elise" activity.*

A lead sheet includes both melody and chords and serves as the foundation for most improvisational activities. In this activity, you will input the melody and chords to Beethoven's "Für Elise" with some slight modifications by yours truly.

TIP

The *Classical Music Fakebook* published by Hal Leonard contains over 850 classical music hits in lead sheet format. Visit this link for more information: http://www.halleonard.com/product/viewproduct.action?itemid=240044&subsiteid=7&.

When you are improvising over "Für Elise," the A sections center on the A harmonic minor scale while the B section uses a combination of C major scale, G major pentatonic scale, A harmonic minor scale, and F major scale. NOTE: Fmaj7 is not typically performed in "Für Elise"; I added it as a slight reharmonization of the B section.

TIP

For beginning improvisers, turn the improvisational sections of this tune into a 2 bar Am-E7 vamp and instruct students to restrict their improvisation to the notes of the A harmonic minor scale.

Examine the lead sheet for "Für Elise" in Figure 3.1. Notice that there is a good bit of repetition throughout the tune.

Figure 3.1
"Für Elise"
(by Beethoven,
arr. Fein) lead sheet.

TIP

Recognizing repetition will save you a tremendous amount of time when you are entering notation because you can use copy/paste.

TIP

Note Flight currently has two views available (HTML5 and Flash). I suggest using HTML5 view only since this view is common to iPad and all other devices including Mac/Windows computers. To switch views, select *Action > View > Switch to HTML5 Editor* in the Note Flight window.

TIP

In the newest Note Flight Score Editor interface, most menus are accessible through the *Action* icon ≡ in the upper left corner of the interface.

TIP

Visit https://www.noteflight.com/guide for the full Note Flight user guide.

1. Login to your Note Flight account.
2. Click *New Score* in the upper left corner of the screen and choose *Start from a blank score sheet* when prompted.
3. Enter the title and composer by clicking the grayed out *(Title)* and *(Composer)* sections.
4. To turn the default grand staff into a treble staff only, double click the lines/spaces of the bass clef staff to select all of the measures in that staff and select *Action > Edit > Delete.*
5. Select the entire score (*Actions > Edit > Select All*) and set the time signature by selecting *Action > Measure > Change Time Signature.* Set the beat count to 3.
6. Deselect the measures by clicking anywhere in the white portion of the screen.
7. Since there is a pickup in this score, select bar 1 only (click the line/spaces of bar 1) and select *Action > Measure > Change Time Signature* again. This time set the beat count to 1 and click the button for *Pickup.*

You are now ready to begin entering the notes and chords. We will go back and add the repeats, 1st/2nd ending, rehearsal letters, and double bars later.

TIP

To enter pitches in Note Flight, simply type the name of the pitch on your QWERTY keyboard. Then set the rhythmic duration and add any accidentals AFTER you have entered the pitch. To enter a rest, tap the spacebar to create a rest and then select the rhythmic duration.

TIP

Use your up/down arrow keys to move a pitch up or down on the staff or press hold CTRL (Windows) or Command (Mac) along with the up/down arrow keys to move a pitch up or down an octave.

8. Enter the notation for bars 1–4.
9. Click the lines/spaces in bar 1 (skipping the pickup measure), hold SHIFT, and click the lines/spaces of bar 4 to select this four-bar phrase.
10. Select *Action > Edit > Copy.*
11. Select bar 4 and select *Action > Edit > Paste.*

12. Correct the pitches/rhythms in bar 7/8.
13. Select the lines/spaces of bar 8 and select *Action > Measure > Add Measure After* to add a bar.
14. Copy bar 1 and paste it into bar 9.
15. Continue to enter the notation for bars 10/11 (2nd ending) and the bridge (bars 12–18).

TIP

Enter the notation for bar 12 and then press letter R (or select *Action > Edit > Repeat*) twice and change the pitches for bars 13/14. Enter the notation for bar 15 and then press letter R three times and change pitches/octaves for bars 16–18.

16. Enter the chords for the A and B sections of the song by clicking a note and pressing letter K on your QWERTY keyboard. You can also click the *Action > Text > Chord Symbol.*

TIP

To enter chords very quickly using your QWERTY keyboard, type the chord name, press RETURN, and press the right facing arrow until you get to the next pitch with a chord symbol. Then tap K and repeat the process.

17. Copy bars 1–8 and paste to bars 20–27. Notice that we waited to copy/paste the last 8 bars until after the chords were already entered to save time.
18. Select bar 3. Select *Action > Text > Rehearsal Letter.*
19. Repeat the previous step at bars 12 and 20 to add additional rehearsal letters.
20. Select bar 8 and select *Action > Repeat > Repeat Ending* to add a 1st Ending.
21. Select bar 10 and select *Action > Repeat > Repeat Ending* to add a 2nd Ending. You can change the "1." to a "2." to indicate the 2nd ending by double-clicking the ending number.

TIP

You may need to click and drag the *Am* chord symbol to move it if the chord symbol is running into the "1." or "2." in the endings.

22. Select bar 3 and select *Action* > *Repeat* > *Start Repeat* to add a start repeat barline.
23. Select bar 9 and select *Action* > *Repeat* > *End Repeat* to add an end repeat barline.
24. Select bar 11 and select *Action* > *Measure* > *Double Bar* to add a double bar.
25. Select bar 19 and select *Action* > *Measure* > *Double Bar* to add another double bar.
26. Correct the layout by selecting the barline at the start of bars 12, 16, and 20 and pressing RETURN (or select *Action* > *Layout* > *System Break*).
27. Finally, click and drag to move items that are colliding such as chord symbols running into rehearsal letters.
28. Save your work by clicking the *Save* icon at the top!
29. Print the lead sheet by selecting *Action* > *Score* > *Print Score* or use the shortcut COMMAND+P (Mac) or CTRL+P (Windows).

TIP

You can also export a PDF file directly from Note Flight. Select *Action* > *Score* > *Export*. Select *Export to PDF* in the pop-up window. NOTE: You must have a paid premium membership to use *Export Individual Parts to PDF*; however, there is a work-around described in a tip below.

30. To prepare a duplicate lead sheet for a transposing instrument:
    a. Click the *Instrument* icon at the top.
    b. Click *Add Instrument* in the pop-up window and select any instrument from the list.
    c. Click the X in the upper right corner to exit the *Instruments* window.
    d. Double-click to the left of the treble staff to select everything on that staff.
    e. Select *Action* > *Edit* > *Copy*.
    f. Click the first bar of the new instrument staff and select *Action* > *Edit* > *Paste*.
    g. Select *Action* > *Score* and uncheck *Show in Concert Pitch*.
    h. Adjust the octave to suit the range of the instrument staff that you added.

TIP

Although *Export Individual Parts to PDF* (from the *Action* > *Score* > *Export* menu) is disabled in the free version of Note Flight, you can actually print individual parts one at a time from the *Perform* area or save them as a PDF. Click the *Perform* icon to enter *Perform* mode. Set the *Perform* menu at the top of the interface to the part you'd like to print or save to PDF. You now can set the *Show* menu also at the top of the interface to "Your Instrument." See Figure 3.2 for an example of these settings.

TIP *continued*

Figure 3.2
Example settings for printing an individual part or saving it to PDF in the *Perform* mode of Note Flight.

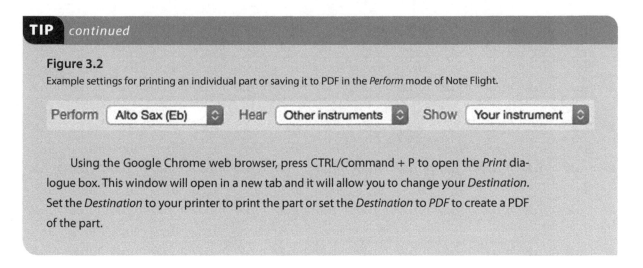

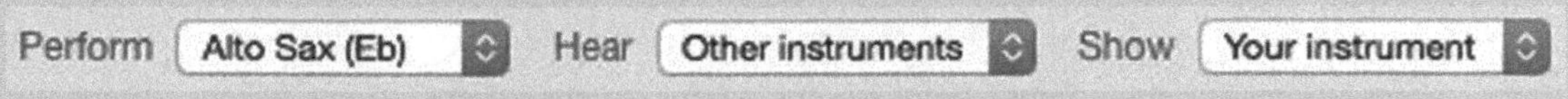

Using the Google Chrome web browser, press CTRL/Command + P to open the *Print* dialogue box. This window will open in a new tab and it will allow you to change your *Destination*. Set the *Destination* to your printer to print the part or set the *Destination* to *PDF* to create a PDF of the part.

Creating an Exercise Sheet: "Money" by Pink Floyd

[▶] *See Example 3.2 on the companion website for XML and PDF example files for the "Money" activity.*

The blues form heavily influenced rock music. Pink Floyd follows a basic blues form in their song "Money" from the acclaimed album *Dark Side of the Moon*. Released in 1973, this landmark album has sold over 50 million copies and currently still sells thousands of copies each week. "Money" was one of two singles released on this album. The catchy bass line in 7/4 meter drives the work throughout the introduction, vocal sections, and improvised saxophone solo while following a basic blues structure. Notice that Pink Floyd does not modulate to the IV chord at the start of the second phrase and instead stays on the I chord for a full 8 bars before moving to the V, IV, I turnaround during the third phrase (see Figure 3.3).

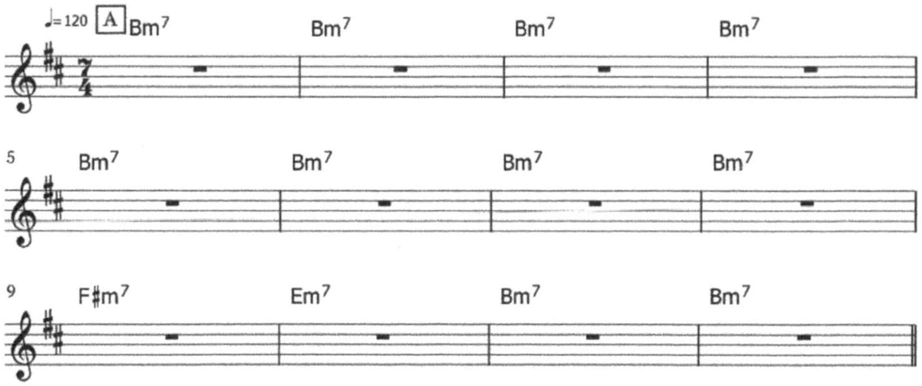

Figure 3.3
Chord progression for Pink Floyd's "Money" throughout the 7/4 sections.

The tune shifts to common time (4/4 meter) during the guitar solo sections and the chord structure changes slightly. Notice that the IV chord now appears at the start of the second phrase but is absent in the middle of the third phrase as the full band plays the unison descending line (see Figure 3.4).

Figure 3.4
Chord progression for Pink Floyd's "Money" throughout the common time (4/4) section.

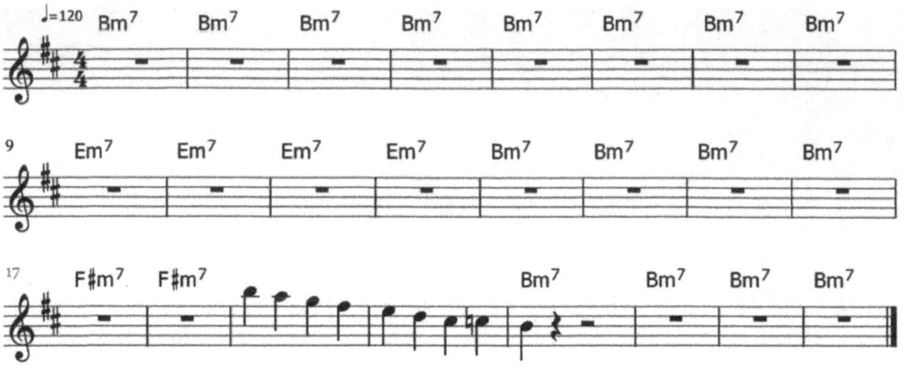

Listen to "Money" on iTunes:

https://itunes.apple.com/us/album/money/id1065973699?i=1065973708.

To develop improvisational flexibility, I recommend creating various patterns for students to practice. In my experience, beginning improvisers need these patterns written out over the entire form and in each key of the tune while intermediate/advanced improvisers can usually handle receiving a page with a single instance of the pattern in each key.

Here's an example of a 1-3-5-7 exercise I would provide a beginning improviser for "Money" (see Figure 3.5).

Figure 3.5
1-3-5-7 exercise for "Money" appropriate for beginning improvisers.

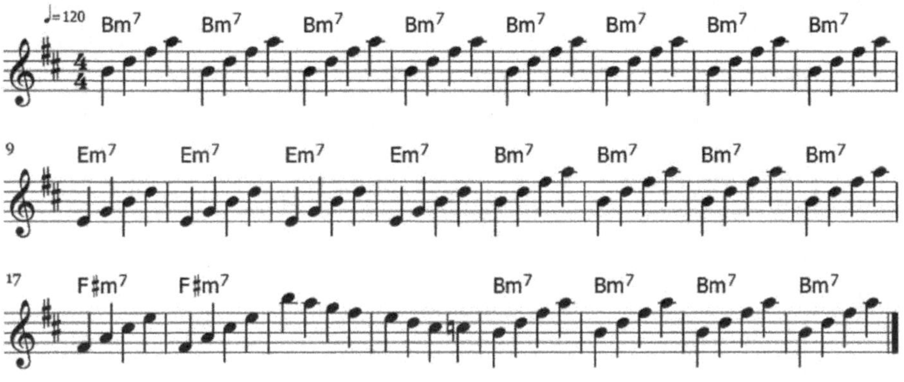

Here's an example of the same exercise more appropriate for intermediate/advanced improvisers for "Money" (see Figure 3.6).

Figure 3.6
1-3-5-7 exercise for "Money" appropriate for intermediate/advanced improvisers.

Let's create a series of exercises to develop facility over the chords to "Money." For brevity's sake, I will follow the intermediate/advanced model of providing a single instance of each exercise in each of the three keys in this tune.

1. Create a new score in Note Flight.
2. Enter a title (Money) and composer (Pink Floyd arr. Your_Name) for the score.
3. Select all measures and select *Action > Measure > Change Key Signature.* Set the key to B minor.
4. For the first pattern, enter 7-5-3-1 of each chord in quarter notes (see Figure 3.7).

Figure 3.7
7-5-3-1 of each chord in quarter notes for "Money."

> **TIP**
>
> I suggest entering the pattern once in Bm and then tap letter R to repeat the pattern. Select *Action > Pitch > Transpose* to move the pattern up a P4 to Em and a P5 to F♯m. Notice that the chords also transpose!

5. For the second pattern, enter 5-7-1-3-1 of each chord to mimic the end of the 7/4 bass line (see Figure 3.8).

Figure 3.8
5-7-1-3-1 of each chord in "Money" to mimic the end of the 7/4 bass line.

6. For the third pattern, enter 5-3-4-2-3-1-7 of each chord to create a difficult twister pattern (see Figure 3.9).

Figure 3.9
5-3-4-2-3-1-7 of each chord in "Money."

7. Create a variety of new instrument staves for each instrument in your ensemble or lesson studio.
8. Copy and paste each exercise into these new staves. You will likely need to transpose sections up or down to fit the range of the specific instrument(s).

9. Click the *Perform* icon and print each part as described in the "Für Elise" activity earlier in this chapter.

TIP

For additional scale patterns explore the following resources:

- Jazz Handbook by Jamey Aebersold available FREE online: http://www.jazzbooks.com/mm5/merchant.mvc?Screen=FQBK
  - Explore p. 10—*Practice Procedure for Memorizing Scales and Chords to Any Song.*
- Patterns for Jazz by Jerry Coker
  - Treble Clef Edition: http://www.alfred.com/Products/Patterns-for-Jazz-A-Theory-Text-for-Jazz-Composition-and-Improvisation--00-SB1.aspx
  - Bass Clef Edition: http://www.alfred.com/Products/Patterns-for-Jazz-A-Theory-Text-for-Jazz-Composition-and-Improvisation--00-SB72.aspx

Creating a Motif-Based Solo: "Four" by Miles Davis

[▶] *See Example 3.3 on the companion website for XML and PDF files for the "Four" activity.*

Melodic motifs are the lingo of music. Each genre has typical motifs (or *licks* as we call them in jazz) and these melodic fragments help improvisers develop the language of the genre. In this activity, you will develop a solo based on some standard jazz licks for "Four" by Miles Davis that fit over the IMaj^7 chord or the ii^7-V^7 chords. I selected "Four" for this activity because it follows a ii-V-I format.

Listen to "Four" on iTunes:

- Miles Davis: https://itunes.apple.com/us/album/four/id182847588?i=182847601
- Dexter Gordon: https://itunes.apple.com/us/album/four/id159111255?i=159111505

Examine these jazz licks that work over ii^7-V^7 chords (see Figure 3.10).

Figure 3.10 Sample jazz licks that work over ii^7-V^7 chords.

Examine these jazz licks that work over IMaj^7 chords (see Figure 3.11).

Figure 3.11

Sample jazz licks that work over IMaj^7 chords.

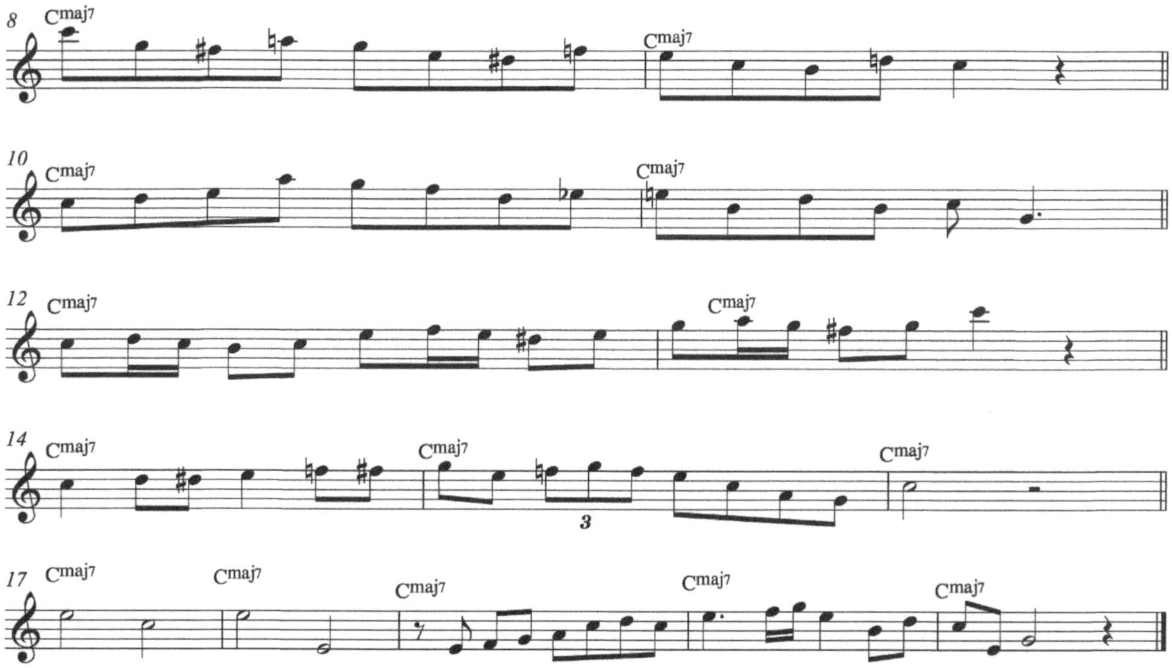

TIP

Discover hundreds more licks from the following sources:

- *How to Play Bebop,* Vol. 2 by David Baker.
- *Charlie Parker Omnibook* transcribed by Jamey Aebersold.
- You can also create your own book of motifs! Select any song in which you know the chord changes. Listen to a recording of a great improviser playing the song. Pause the recording and transcribe the melodic motif along with the sounding chord(s).

1. Visit the companion website to this book.
2. Download the files *jazz_licks.xml* and *motif_solo_template.xml.*
3. Login to your Note Flight account and create a new score. Instead of starting with a blank score, select *Start by importing Music XML or MIDI files > Choose file.*
4. Select the *jazz_licks.xml* file you downloaded.

TIP

Music XML files allow users to migrate between various notation software titles. For example, saving as Music XML would allow a Sibelius user to send a file to a Finale or Note Flight user while maintaining most formatting. To save as Music XML in Note Flight select *Action > Score > Export* and choose *Music XML Score.*

TIP

After you select the xml file, check all the boxes under the filename to maintain as much of the original notation file formatting as possible.

5. Create a new tab in your web browser, go to your Note Flight account, and create another new score. This time import the *motif_solo_template.xml* file you downloaded.

Throughout these next steps, you will select and copy a measure or group of measures from the *Jazz Licks* file and paste it into the *Motif Solo Template* file.

6. Copy and paste a variety of Maj^7 motifs for the Maj^7 chords in the piece. Follow these transposition guidelines to maintain the correct chord progression:
   a. E♭Maj^7 (bars 1–2, 17–18, and 31–32): Up m3 or down M6
   b. A♭Maj^7 (bars 5–6, 21–22): Up m6 or down M3
7. Copy and paste a variety of ii-V7 motifs for the rest of the piece. Follow these transposition guidelines to maintain the correct chord progression:
   a. E♭m^7-A♭^{7} (bars 3–4, 11–12, 15–16, 27–28, and 30): Up m2.
   b. A♭m^7-D♭^{7} (bars 7–8, 23–24): Up dim5 or down aug4.
   c. Gm^7-C^7 (bars 9, 13, 25, 29.1–29.2): Up P4 or down P5.
   d. F♯m^7-B^7 (bars 10, 14, 26, 29.3–29.4): Up M3 or down m6.
8. Play the file using the computer audio output and evaluate the solo.
   a. Do the notes generally flow smoothly from measure to measure? If not, you may consider transposing a given lick up or down an octave to move more smoothly to the next measure. You typically don't want to have big leaps between licks if possible.
   b. Did you transpose the licks accurately? Compare the chord changes of your completed file with the chord changes from the original *Jazz Solo Template* file. Note: In the measures that were previously ii^7 for a full measure followed by V^7 for a full measure, you will now see ii^7 for 2 beats and V^7 for 2 beats and this is OK.
9. Create a variety of new instrument staves for each instrument in your ensemble or lesson studio.
10. Copy and paste the entire motif-based solo into these new staves. You will likely need to transpose sections up or down to fit the range of the specific instrument(s).

TIP

You may need to change the enharmonic spelling of a note for ease of reading. Although theoretically correct, double flats, double sharps, C♭, B♯, Fb, and E♯ are difficult for students to accurately sight-read. In Note Flight, select a note and tap letter Z (or select *Action* > *Pitch* > *Enharmonic Shift*).

11. Click the *Perform* icon and print each part as described in the "Für Elise"activity earlier in this chapter.

See Figure 3.12 for an example of a motif-based solo.

Figure 3.12

Example of a motif-based solo over "Four."

Creating a Chord-Tone Solo: "All the Things You Are" by Kern/Hammerstein

[▶] *See Example 3.4 on the companion website for XML and PDF files for the "All the Things You Are" activity.*

Originally written for the musical *Very Warm May* in 1939 and recorded by such famous artists as Frank Sinatra, Sarah Vaughn, and even Michael Jackson, "All the Things You Are" is truly one of the gems of the Great American Songbook. I particularly enjoy the heavy use of the cycle-of-fourths throughout the song. The form essentially follows the traditional AABA structure but includes a few twists that make this tune unique. Examine A1(see Figure 3.13) and A2 (see Figure 3.14) below and notice that they are identical except that the A2 is transposed up a perfect 5th.

Figure 3.13

Chord progression for the A1 section of "All the Things You Are."

Figure 3.14

Chord progression for the A2 section of "All the Things You Are."

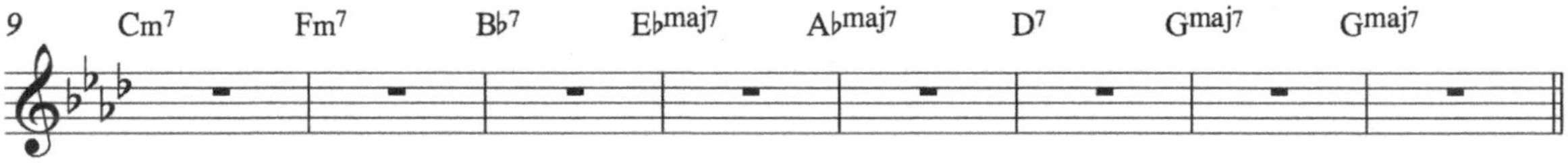

The bridge uses a series of two ii-V7-I progressions (see Figure 3.15).

Figure 3.15

Chord progression for the B section of "All the Things You Are."

The A3 closely follows A1 except for an interesting diversion in the 6th bar of the section that adds 4 additional bars to the A3 (see Figure 3.16).

Figure 3.16

Chord progression for the A3 section of "All the Things You Are."

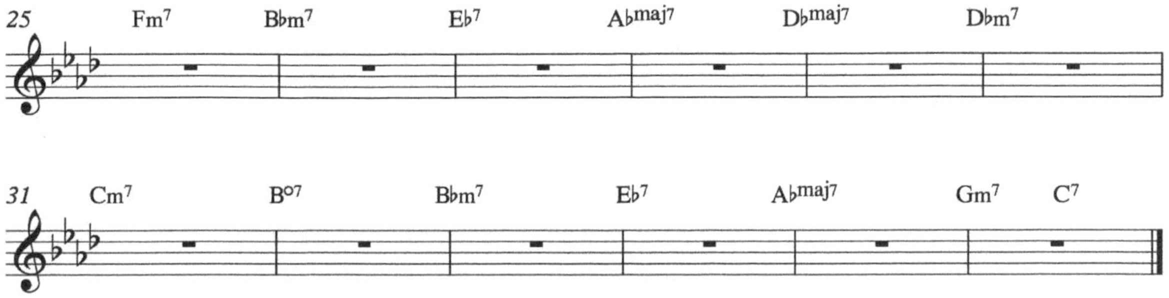

Thinking back to the "tool box" approach to improvisation, we could try a few different "tools" on this tune including chord bracketing and developing a motif-based solo. For this activity, however, I'm going to focus on restricting my improvisation to chord tones only. "All the Things You Are" is especially good for this concept because the guide tones (3rd and 7th of each chord) flow incredibly smoothly with the use of cycle-of-fourths and half-step-away chords. Examine the guide tones for A1 below in Figure 3.17. As you can see, the improviser can generally move from the 3rd of the first chord to the 7th of the next or vice versa.

Figure 3.17

Guide tones for A1 of "All the Things You Are."

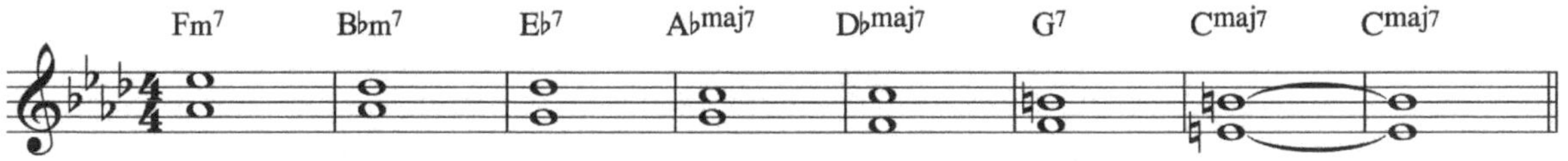

For A1, compose a solo using only chord tones in whole note rhythms (see Figure 3.18).

1. Download the file *chord_tone_solo_template.xml* from the companion website.
2. Create a new Note Flight score and import the XML template file you downloaded.
3. Enter a chord tone for Fmin^7 and set the rhythmic value to whole note.
4. Continue entering chord tones for each measure focusing on stepwise or common tone motion.

Figure 3.18

Example using only chord tones in whole note rhythms over the A1 section of "All the Things You Are."

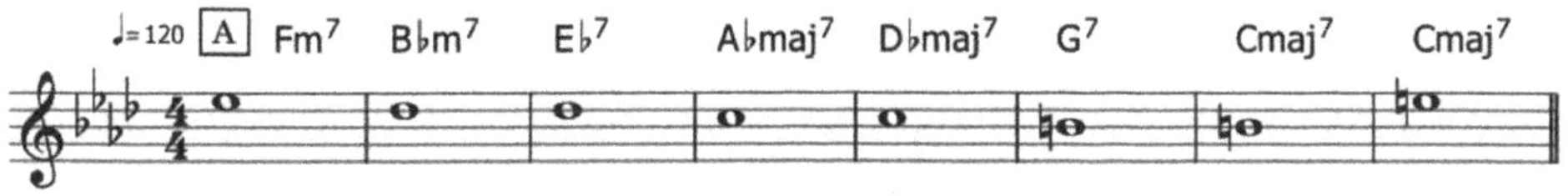

For A2, compose a solo using only chord tones in half note rhythms (see Figure 3.19). Feel free to use leaps of a 5th, 7th, or octave within a measure but it is best to keep your pitches as close together as possible when you move from measure to measure. This allows your ear to begin to hear that smooth transition from chord to chord over the barline.

5. Enter a chord tone for Cmin^7 and set the rhythmic value to half note.
6. Enter another chord tone for Cmin^7 for beats 3–4.
7. As you select your chord tone for Fmin^7 try to select either the same or a nearby note that would fit the Fmin^7 chord. For example, resolve a B♭ in Cmin^7 to an A♭ in Fmin^7.
8. Continue entering chord tones for each measure focusing on stepwise motion or common tones when moving from measure to measure.

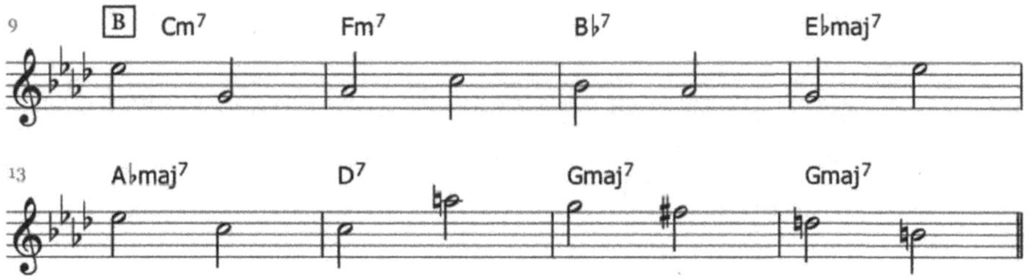

Figure 3.19 Example using only chord tones in half note rhythms over the A2 section of "All the Things You Are."

For the bridge, compose a solo using only chord tones in quarter note rhythms (see Figure 3.20).

9. Select a chord tone for Amin^7 and set the rhythmic value to quarter note.
10. Enter additional chord tones for beats 2, 3, and 4. I suggest using a combination of chord tones ascending/descending in order along with more jagged patterns to keep an interesting melodic contour.
11. Continue entering chord tones for each measure and, again, be sure to keep the pitches close together as you go over the barline to connect the chords smoothly.

Figure 3.20 Example using only chord tones in quarter note rhythms over the bridge (B section) of "All the Things You Are."

Finally, make use of non-chord tones (NCTs) on upbeats to add interest to your chord-tone solo in the A3 section (see Figure 3.21).

TIP

I suggest restricting yourself to the following NCTs.

- Passing tones: These NCTs will ascend/descend within the mode or chromatically. They are generally used to connect notes that are a m2, M2, m3, or M3 apart.
- Neighbor tones: These NCTs will join common tones with a single note above or below. When using neighbor tones above, stay within the mode (diatonic upper neighbor). When using neighbor tones below, use a half-step (chromatic lower neighbor).
- Appoggiatura: These NCTs will join notes that are larger than a third apart. For ascending intervals, leap up and resolve down by step to the next chord tone. For descending intervals, leap down and resolve up by step to the next chord tone.

12. Start with quarter note chord tones only as you did in the bridge section.
13. After you have entered the chord tone downbeats, begin adding in upbeat non-chord tones. Do this by selecting an existing chord tone and changing it to an eighth note and adding the appropriate NCT on the following upbeat. You do NOT need to add NCTs to every upbeat; pick your spots!

Figure 3.21 Example using only chord tones on downbeats and non-chord tones on eighth-note upbeats over the A3 section of "All the Things You Are."

Now that you have completed the activity, take note of the twofold goal of this process. First, students are training their brains to think about chord tones and how to transition from chord to chord smoothly. As you go through this activity there are many constants such as the motion from 3rd to 7th, major 3rd to minor 3rd, and major 7th to minor/dom 7th that directly carry into every improvisational setting. Second, students can use the chord tone solo as a guide during improvisation. The framework of the chord tone solo allows students to modify the rhythm and/or omit notes. It also gives them the courage to venture beyond the given chord tone solo knowing that they can jump right back to the guide if they have trouble.

The Power of MIDI in Notation Software

MIDI is the universal language of all music software. By exporting as a standard MIDI file, users can take musical material created in one application and import it into a completely different application. In this chapter, we export a MIDI file from accompaniment software and import it into notation software so we can see the actual musical material.

1. Open one of the accompaniments you created in Band-in-a-Box in Chapter 2 of this text. I suggest using one of the accompaniments with a soloist track such as "Autumn Leaves" or "So What."
2. In Band-in-a-Box, select *File* > *Save as Standard MIDI File* and select *File on Disk* in the next window.
3. Save the file to a memorable location. Notice that the file extension is *.MID*.

TIP

You can also export MIDI from iReal Pro on your iPad. Select *Share* > *Share Audio* and select MIDI. You can now email your file to yourself for easy access on your computer or save it to Dropbox or Google Drive.

4. In your web browser, login to your Note Flight account.
5. Create a new Note Flight score and select *Start by importing Music XML or MIDI files.*
6. Select *Choose Files* and located the MIDI file you exported previously from Band-in-a-Box (or iReal Pro on your iPad).

TIP

The options you choose after you select your MIDI file will all depend on your specific MIDI file. You will often find yourself importing the same MIDI file multiple times to get the notation looking as good as possible with minimal editing. Here's my suggestion for where to start with these settings:

- *Create Parts From*: You will almost always want to select *Tracks*.
- *Quantization*: Quantization refers to adjusting rhythms to fit in a specified value. In most cases, you must select the smallest rhythmic duration in your MIDI file. If the resulting notation has too many extra rests and short duration notes, try reimporting with a lower quantization value (Ex.: Use 1/8th instead of 1/16th). If the resulting notation has too many notes on top of each other, try reimporting with a higher quantization value (Ex.: Use 1/16th instead of 1/8th). You will almost never select 1/32 for your quantization value.
- NOTE: I typically get the best results by checking both boxes for *Quantize Dotted Rhythms* and *Extend Short Notes through Rests*.

7. Enter the chords for a single chorus above one of the parts such as the Bass staff.
8. Select the first measure with chords, hold the shift key, and select the last measure with chords.
9. Select *Action > Edit > Filter Selection* and choose *Chord Symbols*. Now only the chord symbols are selected. Click the X in the Filter window to close it.
10. Select *Edit > Copy*.
11. Click the first measure of each part and select *Edit > Paste* to apply those same chords to all parts. Continue pasting the chords for each chorus of the tune.
12. Select the first measure of each chorus and add a rehearsal letter (*Action > Text > Rehearsal Letter*). This will help students identify the top of the form for the song.

We are ready to begin using the material for specific purposes. We will start by creating printouts for the soloist track.

13. Select the Soloist staff.
14. Click the *Instrument* icon and add multiple staves for various transposing instruments in your ensemble.
15. Copy the material from the original Soloist staff and paste it into the new instrument staves.
16. Click the *Perform* icon in the upper left corner of the screen.
17. Click the *Parts* icon in the upper left corner of the *Perform* screen.
18. Select one of the new staves you created in step 13. Press CTRL/Command + P to print the part. Repeat with each of the new staves to generate the soloist material for your ensemble instrumentation. Remember, in the Google Chrome Web Browser you can also set the *Destination* in the print dialogue box to *PDF* to save the soloist staff as a PDF file.
19. Click *Done* in the upper left corner to exit the *Perform* screen.

TIP

You will likely need to change octaves for portions of the soloist material based on the range of the instruments in your ensemble. Select the measure(s) that require transposition and select *Action > Pitch > Move Up an Octave* or *Move Down an Octave*.

TIP

To practice performing the soloist material with the accompaniment instruments only select *Parts > Listen to Other Parts*. You will now SEE the soloist notation without hearing a computer-generated performance of the soloist material while at the same time hearing the accompaniment part. This creates a "music minus 1" experience similar to the Aebersold accompaniments discussed in Chapter 1 of this text.

Let's also examine the bass part to help inform a student bass player how to construct a walking bass line.

20. The bass part is most likely an octave too low. Select all of the bass material by double clicking the lines/spaces of any bar in that staff.
21. Press CTRL/Command + ↑ to transpose the material up an octave (or select *Action > Pitch > Move Up an Octave*).
22. Click the *Perform* icon in the upper left corner of the screen.
23. Click the *Parts* icon in the upper left corner of the *Perform* screen.
24. Select bass staff and press CTRL/Command + P to print the part or save it as a PDF.

iPad Connection: forScore

As a gigging musician and high school jazz band director, forScore is my app of choice for loading all of my sheet music into my iPad. The days of oversized folders and binders with paper scores and sheet music are gone. I only carry my iPad with me to gigs, rehearsals, and performances!

Before we navigate forScore, we need music to load into the application.

Exporting a PDF from Note Flight

Because of the variations of various web browsers, I'm going to include the steps for exporting a PDF from Note Flight using the free Google Chrome web browser. Generating PDFs from notation software will create very small file-sized PDFs most ideal for your iPad. Although the steps listed below are mentioned previously throughout this chapter, I have included the PDF export steps again for your convenience.

1. Open any existing Note Flight score.
2. To export the full score:
   a. Select *Action > Score > Export.*
   b. Select *Export to PDF.*
3. To export an individual part, premium members can select *Action > Score > Export* and select *Export Individual Parts to PDF.* Free account users must follow the following work-around steps:
   a. Select *Perform* from the upper left corner of the screen.
   b. Select the material you'd like to print (either an individual part or the entire score).
   c. Press CTRL/Command + P. After you press the *Print* button, a new tab will open with a print dialogue box.
   d. Next to *Destination* select *Change* (see Figure 3.22). Under *Local Destinations* select *Save as PDF.*
   e. Click *Save* and choose a memorable location on your computer.

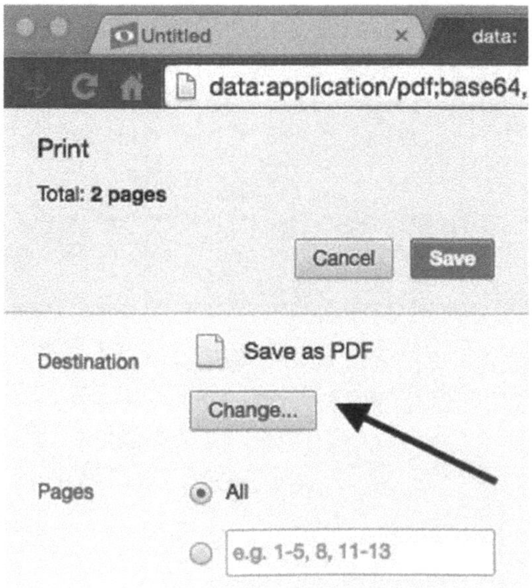

Figure 3.22 In the Google Chrome *Print* dialogue box, select *Change* next to *Destination*. Then, under *Local Destinations* select *Save as PDF*.

Scanning Existing Sheet Music

Although you could take the time to enter the notation into Note Flight, if there is no need for editing or transposition of the music, a simple scan of existing sheet music is often the best solution.

1. Set your scanning software to the following settings:
   a. *Black and white*: Selecting color or grayscale will generate a PDF significantly larger than a black-and-white-only file. Smaller files will load faster and take up less space on your iPad.
   b. *300 dpi*: Higher dpi files will have more detail but take up more disk space. In my experience 300 dpi generally provides enough detail for the staff lines and note stems while keeping file size reasonable. Scores will often require more detail and I will sometimes bump the dpi settings up until I can adequately see the staff lines on each part clearly.
2. Scan the pages and save the PDFs to a memorable location.

TIP

If you are scanning multi-page documents, set your scanning software to "append" or create "multi-page" PDFs. This allows you to create a single multi-page PDF for your sheet music.

TIP

For additional forScore scanning tips, visit http://forscore.co/kb/scanning-tips/.

TIP

You can also take a photograph using your built-in iPad camera in forScore. Since PDFs and black/white scanned music look better and take up less space, I only use the photo option when I'm in a pinch.

To photograph music in the forScore app:

- Tap the *Toolbox* icon in the upper right and select *Darkroom*.
- Tap the *Camera* icon at the top to take a photo of the music using your built-in iPad camera. NOTE: You can also import an existing image from your photo library by tapping the *File* icon next to the *Camera* icon.
- Take a photo of each page of music and then tap *Done*.
- Tap *Save* to create a single score containing all of the images.

Importing PDFs into forScore

Now that you have PDFs of your sheet music, you are ready to import them into forScore.

1. Connect your iPad to your computer using the USB cable.
2. Launch iTunes and select your iPad from the devices area (see Figure 3.23).
3. Select *Apps* from the left pane of the screen. Scroll down to locate *File Sharing* apps.
4. Select forScore in the *File Sharing* list.
5. Click *Add* at the bottom of the *Documents* area and select various PDF files to add them into the forScore app.

Figure 3.23
Launch iTunes and select your iPad from the devices area.

TIP

You can also drag your PDFs into the *Documents* area.

TIP

If you have a Dropbox and/or Google Drive account, install the Dropbox and/or Google Drive apps on your iPad. In forScore, tap the *Toolbox* icon and select *Services*. Then tap + in the upper left corner, select *Dropbox* and/or *Google Drive* (see Figure 3.24). The first time you do this you'll need to tap the arrow next to *Dropbox* and/or *Google Drive* in the *Services* window and enter your account information. This allows forScore to access your Dropbox and/or Google Drive files. Tap *Done* to exit this window.

You can now add PDFs, images, and audio stored in Dropbox and/or Google Drive to your forScore app!

Figure 3.24
Add Dropbox and/or Google Drive to the *Services* area of forScore to all allow forScore access to your Dropbox and/or Google Drive files.

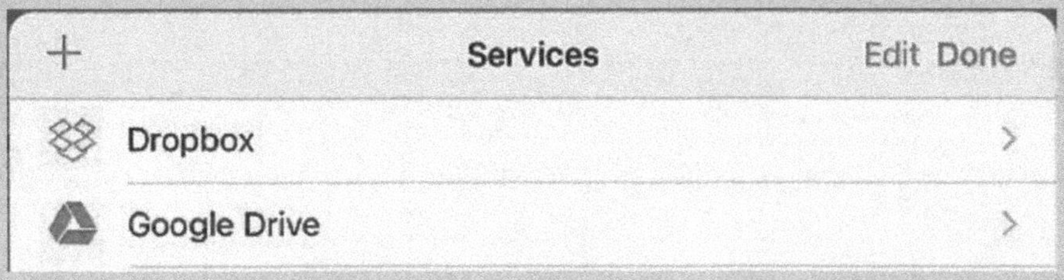

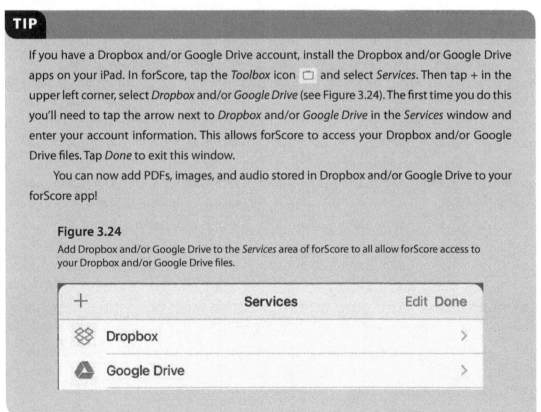

Organizing Music with forScore

You can create set lists with forScore that allow you to easily access your files. I typically place all of my jazz ensemble scores in one set list while I place my jazz lead sheets in a separate set list.

1. To make a new set list, tap the *Setlists* icon ☰.
2. Tap the + icon in the upper right corner of the *Setlists* window and enter a name for your new setlist.
3. Tap the + icon in the new set list.
4. Your files will open up on the right side of the screen. Simply tap any scores/parts you'd like to add to the set list.

TIP

If you have a large multi-page PDF, you can add bookmarks for specific pages. To add a bookmark, open any sheet music file in forScore and tap the *Bookmark* icon 📖. Tap the + icon, enter a name for the bookmark, and tap *Save*. Adding bookmarks will allow you to include a single page or group of pages from a large PDF in a setlist. For example, I have a PDF with about 400 jazz lead sheets. In this large PDF, I bookmark my most commonly played tunes and then I add these tunes to my jazz lead sheet set list.

5. Reorder the files by tapping *Manual* (Manual) under the setlist name. Tap and hold on the 3 lines to the right of the file name and drag up or down to reorder the list.
6. Tap *Done* to finish adding files to the set list.
7. Tap any file in the set list to open it.
8. Tap the right side of the screen to advance to the next page or score in your set list. Tap the left side of the screen to go back a page.

Advanced Features in forScore

1. *Annotations:* This feature allows the user to mark up any score in a variety of ways (see Figure 3.25).

Figure 3.25
Annotations menu in forScore (stamps/shapes, pens, and text).

a. Tap the screen, tap the *Toolbox* icon, and select *Annotate.*
b. *Draw using the various pens:* Tap one of the pen icons and start drawing on the screen to mark up your score. You can customize the pen by tapping the pen icon again. Adjust the pen's hue (color), saturation, transparency, brightness, and size.

   c. *Add stamps or shapes.* The stamps menu includes music notes, accidentals, dynamics, articulations, etc., while the shapes menu includes slurs, brackets, hairpins, and even a 5-line music staff!
   d. *Type in text* to add custom text elements to your score.
   e. Tap *Done* to exit the *Annotate* mode.
2. *Record your performance:*
   a. Select *Toolbox* > *Record* to open the record window.
   b. Tap the *Microphone* icon to begin recording.
   c. Tap the 3 lines next to the *Microphone icon* to access saved recordings for the score. You can also share or delete recordings from this menu.
   d. Tap the X to exit *Record* mode.
3. *Add audio accompaniment:* This is one of my favorite forScore features! Add music onto your iPad in iTunes and then embed play-along accompaniments with scores in forScore so you can view the sheet music while the play-along provides the backing tracks.
   a. Create a new playlist in iTunes for all of your accompaniment tracks and drag your accompaniment audio files into this playlist. You can add accompaniment tracks from existing methods discussed in Chapter 1 (such as Aebersold play-alongs), auto accompaniment software discussed in Chapter 2 (Band-in-a-Box or iReal Pro), or notation software discussed in this chapter (Note Flight).

TIP

Save your Band-in-a-Box, iReal Pro, and Note Flight files as AUDIO.

- Band-in-a-Box:
  - Select File > Save Song as .m4a Audio
- iReal Pro:
  - Tap the *Share* icon.
  - Choose *Audio* and select *AAC*.
  - Email the file to yourself or save to Dropbox or Google Drive.
- Note Flight:
  - Select *File* > *Export* and choose WAV audio.
  - Import this file into iTunes.
  - Select the WAV file and select *File* > *Create New Version* > *Create AAC version*. This will convert the very large WAV file into a much smaller AAC file.
  - NOTE: You can convert to MP3 if you'd prefer. Select *iTunes* > *Preferences*. Under the *General* tab, click the *Import Settings* button. Next to *Import Using*, select *MP3 encoder*.

   b. Connect your iPad to your computer using the USB cable.
   c. Select your iPad from the devices area in iTunes.

d. Select the *Music* tab and be sure that your iPad is syncing with your accompaniment playlist. This provides access to your accompaniment audio files on your iPad within the forScore app.
e. Launch forScore on your iPad and open one of your sheet music files.
f. Tap the score name in the top center to open the *Metadata* panel.
g. Select the *Audio* tab.
h. Tap the *Add from iTunes* button in the bottom right to select songs from your iTunes library synced to your iPad.
i. Tap the screen to exit the *Metadata* panel.
j. Swipe up from the bottom of the screen to open the playback controls.

4. Sharing scores: Especially as you work with students, it is helpful to make annotations in a score during a rehearsal and email a score with annotations to students for further review after rehearsal.
    a. Open any score, tap the *Toolbox* icon, and select *Share*.
    b. You can choose to share the original PDF or annotated PDF. Note: Avoid the 4SC option because this file is only compatible with the forScore app while PDFs are viewable on just about any device or platform.
    c. Tap *Mail* to email the original or annotated PDF.

TIP

Instead of sharing via email, try sharing via Dropbox.

- Instead of tapping *Mail* in the steps above, select *Copy to Dropbox.*
- Your iPad will open the Dropbox app. Tap *Save* to add the file to Dropbox.
- Still in the Dropbox app, tap the down facing arrow next to the file name of the PDF you pulled from forScore, and select *Send Link.*
- Tap *Mail* to email the link to students. Now students can access the PDF without the need to attach the file.
    - Alternatively, you could copy and paste the Dropbox link to the class website. I further discuss web design in Chapter 7 of this text.

TIP

Instead of sharing via email, try sharing via Google Drive.

- Instead of tapping *Mail* in the steps above, select *Google Drive.*
- In the next window tap *Upload* to add the file to your Google Drive account.
- Refer to *Posting and Sharing with Google Drive* in Chapter 6 of this text for information and tips on sharing this PDF with students.

TIP

As with most iOS apps, updates and additions are constantly occurring. For additional forScore support visit http://forscore.co/kb/. Here you will find a detailed user guide, troubleshooting, and FAQs.

Chapter 3 Review

1. In this chapter, we created a lead sheet for "Für Elise." Using the *Classical Music Fakebook* published by Hal Leonard select a tune that would be appropriate for young improvisers. Develop a lead sheet and enter the notation into a new Note Flight score. Consider creating a vamp section for improvising.
2. Examine the lead sheet arrangement of "Fur Elise" used in this chapter. Use Note Flight to create a notation sheet that includes the 4 parent scales used (A harmonic minor, C major, G major pentatonic, and F major) along with a pattern (such as 3rds or a returning scale pattern) for students to practice. Transpose the score for various instruments and print or save as a PDF.
3. Pink Floyd's "Money" follows a 12-bar blues form. Select another blues-based song and create an exercise pattern notation sheet. Some example blues songs include "Watermelon Man" by Herbie Hancock, "I Feel Good" by James Brown, and "Straight, No Chaser" by Thelonius Monk.
4. Using the motifs presented in this chapter, create a motif-based solo over a different ii-V-I based tune such as "Satin Doll" by Duke Ellington.
5. Chapter 2 closely examines "Autumn Leaves." Create a chord tone solo over the chord progression to "Autumn Leaves."
6. When importing a MIDI file, setting the quantization value is very important. Describe quantization and how to appropriately set the quantization value.
7. Keyboard shortcuts can be very helpful when entering notation in Note Flight. Explore the full set of keyboard shortcuts available in Note Flight (http://assets1.noteflight.com/docs/noteflight-keys.pdf). List the shortcuts you frequently use in Note Flight.
8. Describe the benefits of using forScore on the iPad for reading sheet music.
9. Identify the various ways to get sheet music into forScore and discuss the pros and cons of each method.

MUSIC PRODUCTION SOFTWARE 4

Audacity, GarageBand, and GarageBand for iOS

Music production software such as Audacity and GarageBand allow for the manipulation and recording of audio. These applications have become increasingly affordable and accessible for students and teachers over the past 10 years. Audacity, a free and cross-platform application, allows a user to import an existing recording and easily cut and fade sections of audio. This chapter will walk you through the process of isolating an improvised solo and various versions of a melody in existing recordings. GarageBand allows users to work with MIDI as well as audio in a multi-track environment. Specifically, users can export MIDI from iReal Pro or Band-in-a-Box and obtain increased editing control and better software instrument sounds using GarageBand. I discuss the basics of recording acoustic instruments into GarageBand and offer specific uses for adding in acoustic sound. Finally, in the iPad Connection portion of the chapter, I explore GarageBand for iOS. This app is an excellent tool for creating accompaniments mostly because of the included Smart Instruments. Users can simply enter the chords to a song and make use of those chords in a variety of Smart Instruments including guitar, bass, keyboard, and strings.

Isolating and Slowing Down Audio in Audacity: Miles Davis solo on "So What"

[▶] *See Example 4.1 on the companion website for an audio recording example of the the "Miles Davis Solo on 'So What'" activity.*

Transcribing an improvised solo allows the developing improviser to delve deeply into the playing of great artists and analyze his or her improvised idea compared to the sounding chord. Throughout my college years and beyond I have transcribed dozens of solos by my favorite artists including Dexter Gordon, Sonny Stitt, Julian "Cannonball" Adderely, and Paul Desmond. Each time I transcribe, practice, and memorize one of these improvised solos, I take a little piece of that artist with me. It is really the combination of these artists plus my own practice exercises that create my style.

The process of transcribing an improvised solo can be very difficult for young musicians. I remember struggling through 32 bars of a relatively simple Sonny Rollins solo ("Moritat" from *Saxophone Colossus* by Sonny Rollins) when I was a freshman in high school. My ears just weren't able to accurately capture and notate Sonny's performance without assistance from my saxophone teacher. Today, students can take advantage of features in Audacity to isolate a specific section of an audio recording and then slow it down without affecting the pitch.

For this activity, we will use Miles Davis's iconic trumpet solo on "So What" from the album *Kind of Blue*. Refer to Chapter 2 for a detailed analysis of "So What."

- Listen to and purchase "So What" on iTunes: https://itunes.apple.com/us/album/so-what/id268443092?i=268443097

TIP

If you already own "So What" on CD, you can import the CD track into iTunes for use in this activity.

1. Download and install Audacity.
   a. Visit http://www.audacityteam.org/download/ to download the Audacity application.

TIP

In addition to the Audacity application, I would also suggest installing the LAME MP3 Encoder and the FFMPEG Import/Export Library. FFMPEG allows you to import AAC audio files purchased from iTunes while LAME allows you to export as MP3. Explore all of the optional plug-ins in the *Plug-ins* section of the *Downloads* page (http://www.audacityteam.org/download/plug-ins/).

2. Launch Audacity. A new file should automatically open. If a new file does not open, select *File > New*.
3. Select *File > Import Audio*. Locate the Miles Davis "So What" recording to import it into Audacity.
4. Click on the arrow next to the track name and select *Name* to rename the track.
5. Type "Miles NORM" in the *Track Name* dialogue box.
6. Zoom in horizontally on the audio by pressing CTRL/Command + 1 or select *View > Zoom In*.
7. Click your cursor at approximately 1:31 when Miles Davis begins his solo.
8. Select *Edit > Clip Boundaries > Split*.
9. Click your cursor at approximately 3:26 when Miles Davis concludes his solo.
10. Again, select *Edit > Clip Boundaries > Split*.
11. Select *View > Fit in Window* to see the entire file (see Figure 4.1).

Figure 4.1
Split the start and end of Miles Davis's solo in "So What."

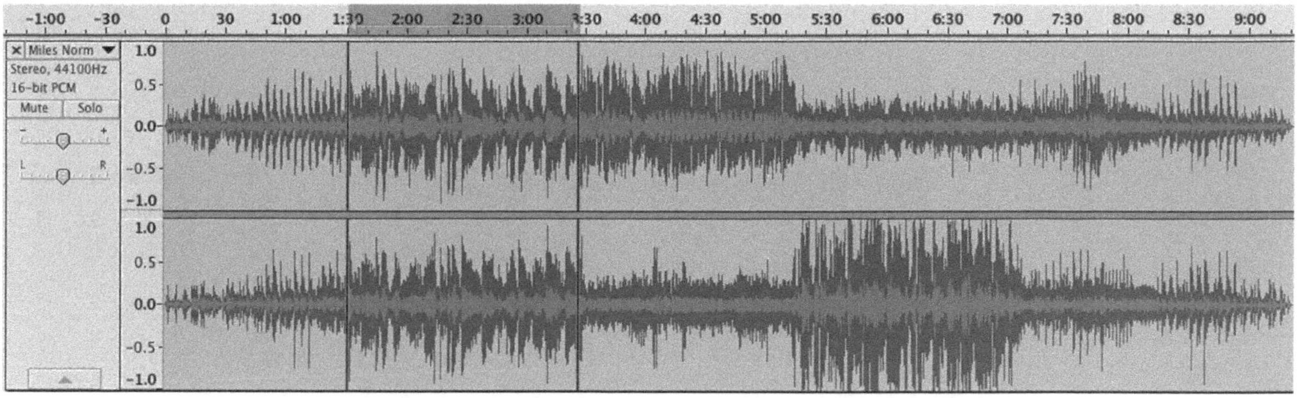

12. Double click the material before your first split to select it and then press the DELETE key on your keyboard. Do the same with the material after your second split. You should now be left with only Miles Davis's solo.
13. Again, select *View > Fit in Window* see what's left of the entire file (see Figure 4.2).

Figure 4.2
Remove all audio material except for Miles Davis's solo in "So What."

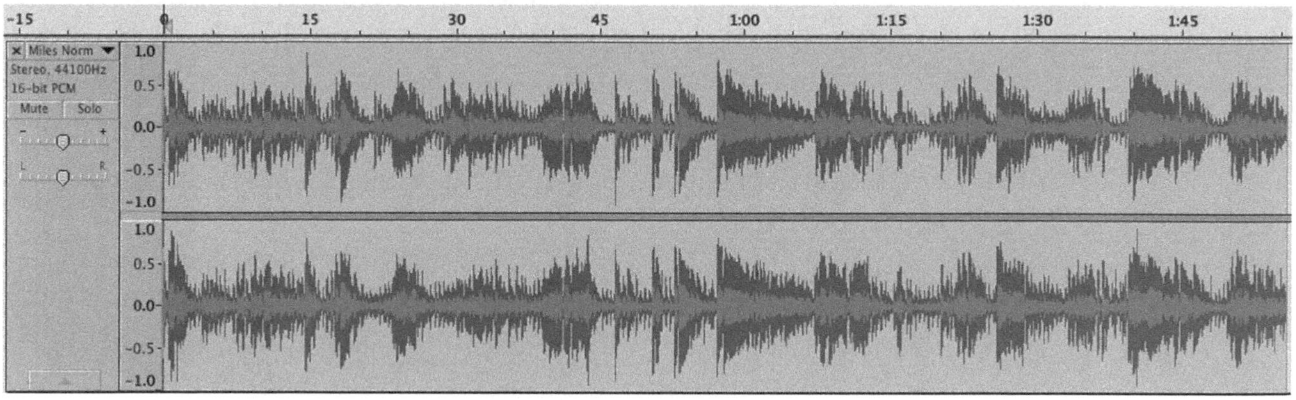

14. Click and drag to select approximately 1 second at the start of the solo. Select *Effect > Fade In* to smooth out the start of the solo section.
15. Click and drag to select approximately 1 second at the end of the solo. Select *Effect > Fade Out* to smooth out the end of the solo section.
16. Double click the audio in the first track to select all of it.
17. Select *Edit > Copy*.
18. Select *Tracks > Add New > Stereo Track*.
19. Name the track "Miles SLOW."
20. Click your cursor at the start of the *Miles SLOW* track and select *Edit > Paste*. You now have two copies of same material.
21. With the audio on the *Miles SLOW* track highlighted, select *Effect > Change Tempo*.

22. Set the percentage change to –25% and click *OK* to process the audio.
23. Select *View > Fit in Window* to fit your entire file horizontally in window (see Figure 4.3).

Figure 4.3

Create a slower version of Miles Davis's solo in "So What."

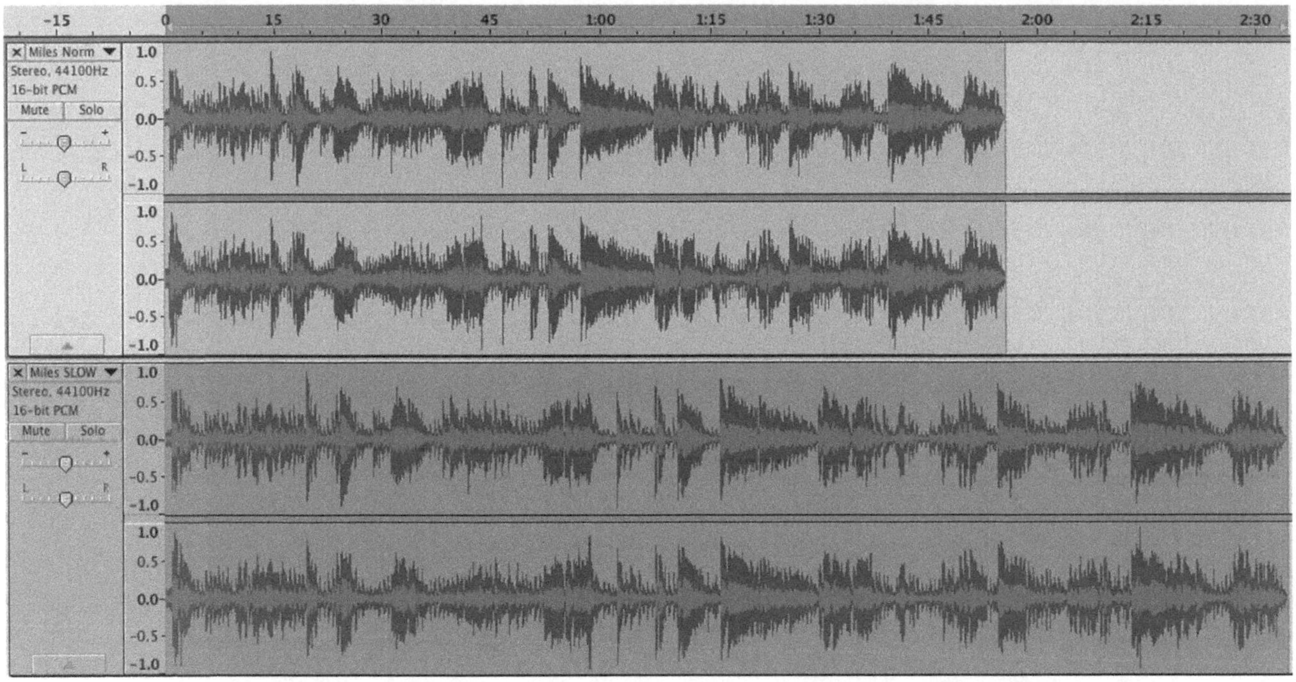

24. Listen to the slower audio in *Miles SLOW* compared to the normal tempo of *Miles NORM.*

TIP

Click *Solo* Solo on the track to isolate a single track. Alternatively, you can click *Mute* Mute on any tracks you don't want to hear.

25. Select *File > Export Multiple.* See Figure 4.4 for an image of the suggested settings.
    a. Set the *Export Format* to <u>*MP3*</u>.
    b. Select a memorable location for the *Export Location* such as your desktop.
    c. Set *Split Files Based On:* <u>*TRACKS*</u>.
    d. Set *Name Files:* <u>*Using*</u> <u>*Label/Track*</u> <u>*Name*</u>.
    e. Click *Export.*
    f. Enter the desired information into the *Edit Metadata* window and click *OK.*

Figure 4.4
Recommended settings in the *Export Multiple* window.

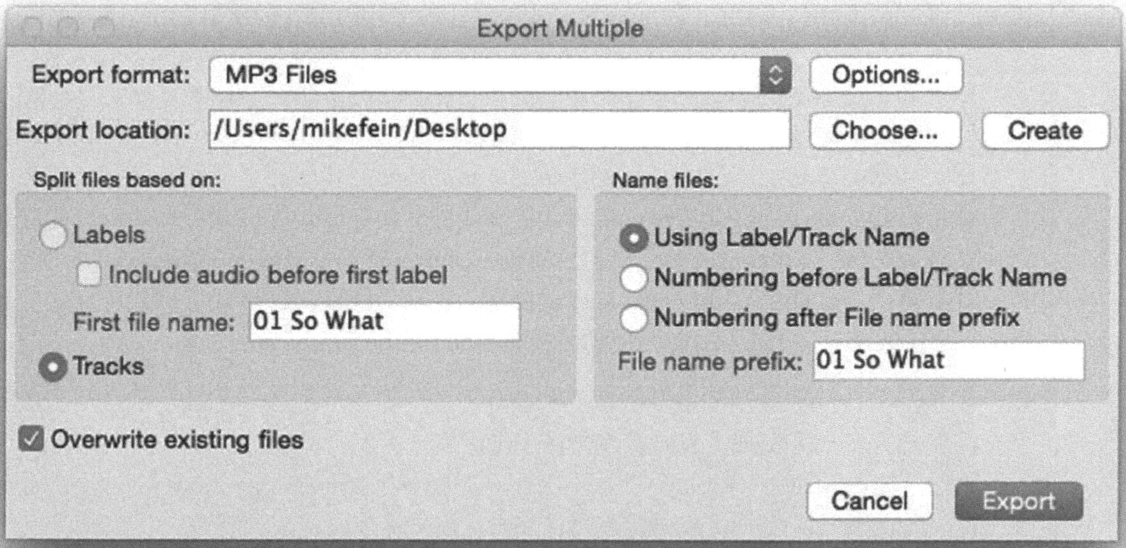

TIP

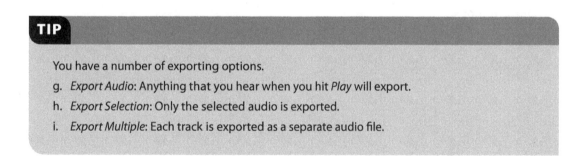

You have a number of exporting options.

g. *Export Audio*: Anything that you hear when you hit *Play* will export.
h. *Export Selection*: Only the selected audio is exported.
i. *Export Multiple*: Each track is exported as a separate audio file.

26. You should now have the normal speed and slowed-down version of the Miles Davis solo saved as MP3 files to your desktop.

TIP

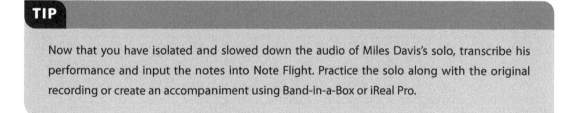

Now that you have isolated and slowed down the audio of Miles Davis's solo, transcribe his performance and input the notes into Note Flight. Practice the solo along with the original recording or create an accompaniment using Band-in-a-Box or iReal Pro.

Comparing Melodies in Audacity: "All Along the Watchtower" by Dylan, Hendrix, and Matthews

[▶] *See Example 4.2 on the companion website for an audio recording example of the the "All Along the Watchtower" activity.*

Learning, ornamenting, and varying melodies is tremendously important. I found that the process of learning a lot of melodies (especially learning melodies by ear only) really helped my improvisational development both by improving my ear and giving me more melodic material in my brain. In this activity we are going to import three versions of "All Along the Watchtower" to compare/contrast the styles of Bob Dylan, Jimi Hendrix, and Dave Matthews. We will focus on the first verse only of each version of this vamp tune.

Listen to and purchase "All Along the Watchtower" on iTunes:

- Bob Dylan - Album: *John Wesley Harding*: https://itunes.apple.com/us/album/all-along-the-watchtower/id181457097?i=181457445
- Jimi Hendrix - Album: *Electric Ladyland*: https://itunes.apple.com/us/album/all-along-the-watchtower/id357652252?i=357653191
- Dave Matthews - Album: *Live at Red Rocks 8.15.95*: https://itunes.apple.com/us/album/all-along-the-watchtower-live/id269532921?i=269534156

"All Along the Watchtower" is in a minor tonality (key of C♯ minor) and only uses the i, VII, VI chords (derived from the C♯ *natural* minor scale). See Figure 4.5 for triads derived from the C♯ natural minor scale.

Figure 4.5
Triads derived from the C♯ natural minor scale.

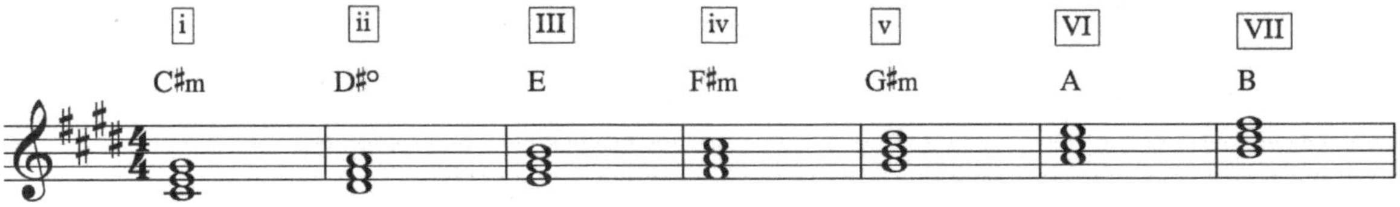

The chords are vamped in the following progression throughout the song (see Figure 4.6):

Figure 4.6
Chord progression for "All Along the Watchtower" (Bob Dylan key).

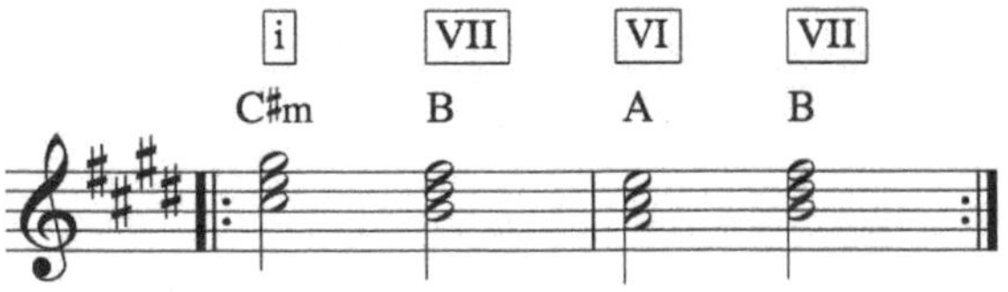

When improvising over this tune it is helpful to consider the bracketing idea discussed over "Autumn Leaves" in Chapter 2. Since all three chords come from the C♯ natural minor scale, the improviser is free to restrict himself or herself to those notes only (see Figure 4.7). Using the C♯ blues scale can add another sonic flavor to the improvisation (see Figure 4.8).

Figure 4.7
C♯ natural minor scale.

Figure 4.8
C♯ blues scale.

Note: Jimi Hendrix performs this tune in C minor while Dave Matthews performed the tune in A minor (see Figure 4.9).

Figure 4.9
Chord progression for "All Along the Watchtower" in C minor (Jimi Hendrix key) and A minor (Dave Matthews key).

1. Launch Audacity. If a new file does not open, select *File > New*.
2. Select *File > Import Audio*. Locate Bob Dylan's version of "All Along the Watchtower" and select *Open*.
3. Click and drag to select the first verse of Bob Dylan's recording (approximately 0:15–0:45).
4. Select *Edit > Clip Boundaries > Split New* to extract the selected audio to a new stereo track.
5. Name the new track "Bob Dylan."
6. Delete the rest of Bob Dylan's original track by pressing the X next to the track name so you are left with only the first verse section on its own track.
7. Select the *Time Shift* tool ↔ from the tool palette.
8. Click and drag Bob Dylan's first verse to the start of the file.
9. Select the *Selection* tool I from the tool palette.
10. Select a small portion of the beginning and ending of the first verse and select *Effect > Fade In/Out* to smoothly fade in/out of the region (see Figure 4.10).

Figure 4.10
Bob Dylan's first verse in Audacity.

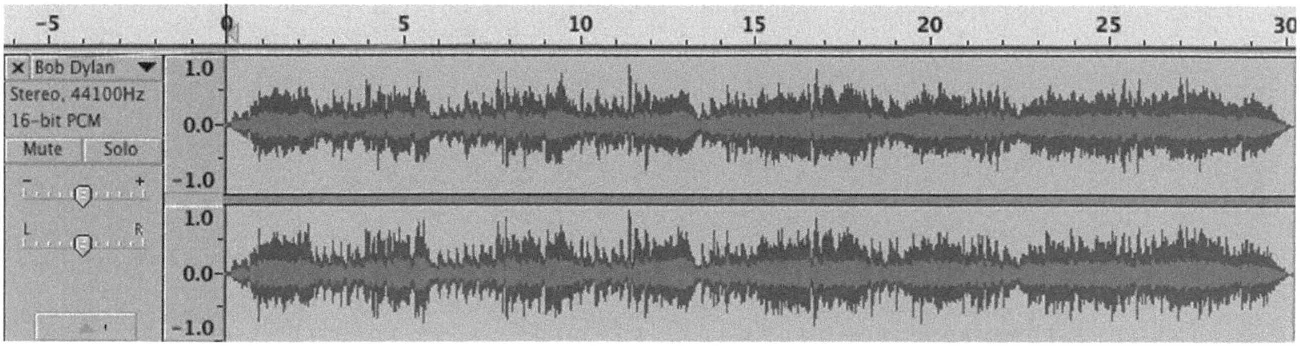

11. Mute Bob Dylan's track for now while we continue to edit additional audio.
12. Select *File > Import Audio*. Locate Jimi Hendrix's version of "All Along the Watchtower" and select *Open*.

13. Click and drag to select the first verse of Jimi Hendrix's recording (approximately 0:18–0:53).
14. Select *Edit > Clip Boundaries > Split New* to extract the selected audio to a new stereo track.
15. Name the new track "Jimi Hendrix."
16. Delete the rest of Jimi Hendrix's original track by pressing the X next to the track name so you are left with only the first verse section on its own track.
17. Select the *Time Shift* tool ↔ from the tool palette.
18. Click and drag Jimi Hendrix's first verse so it immediately follows Bob Dylan's first verse.
19. Select the *Selection* tool I from the tool palette.
20. Select a small portion of the beginning and ending of the first verse and select *Effect > Fade In/Out* to smoothly fade in/out of the region (see Figure 4.11).

Figure 4.11

Jimi Hendrix's first verse in Audacity following Bob Dylan's first verse.

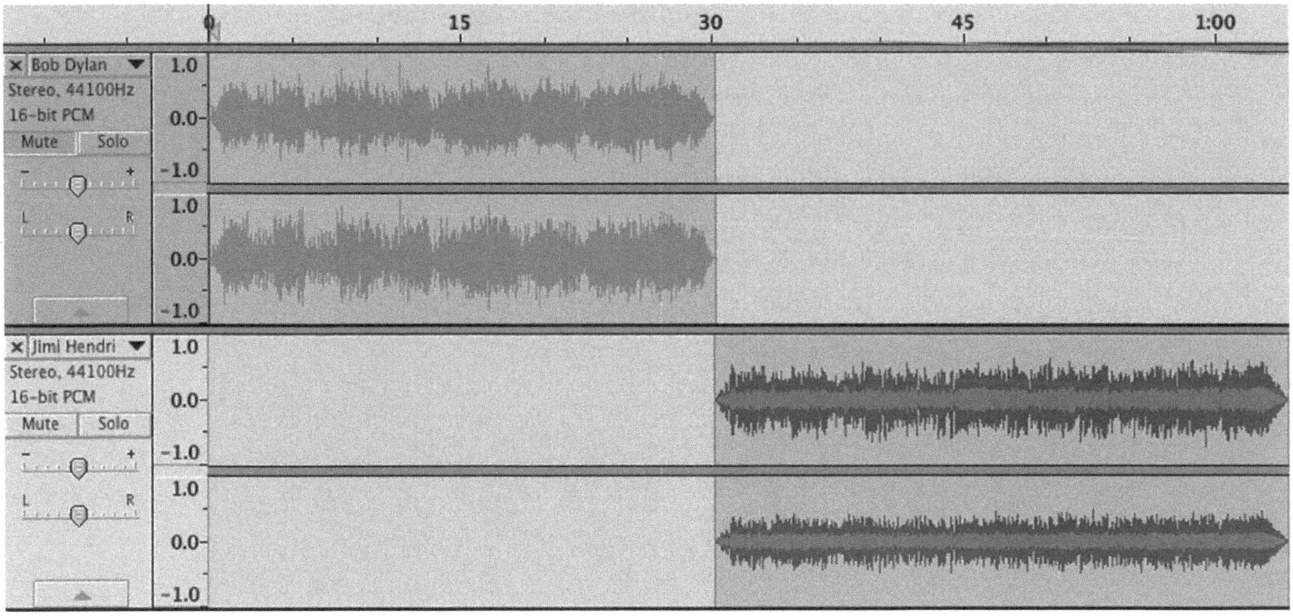

21. Mute Jimi Hendrix's track for now while we continue to edit additional audio.
22. Select *File > Import Audio*. Locate Dave Matthews' version of *All Along the Watchtower* and select *Open*.
23. Click and drag to select the first verse of Dave Matthews' recording (approximately 0:27-2:20).
24. Select *Edit > Clip Boundaries > Split New* to extract the selected audio to a new stereo track.
25. Name the new track "Dave Matthews."

26. Delete the rest of Dave Matthews' original track by pressing the X next to the track name so you are only left with the first verse section on its own track.
27. Select the *Time Shift* tool ↔ from the tool palette.
28. Click and drag Dave Matthews's first verse so it immediately follows Jimi Hendrix's first verse.
29. Select the *Selection* tool I from the tool palette.
30. Select a small portion of the beginning and ending of the first verse and select *Effect > Fade In/Out* to smoothly fade in/out of the region.
31. Unmute all tracks and listen to the entire compilation (see Figure 4.12).

Figure 4.12

Bob Dylan's, Jimi Hendrix's, and Dave Matthews's first verse of "All Along the Watchtower" in Audacity.

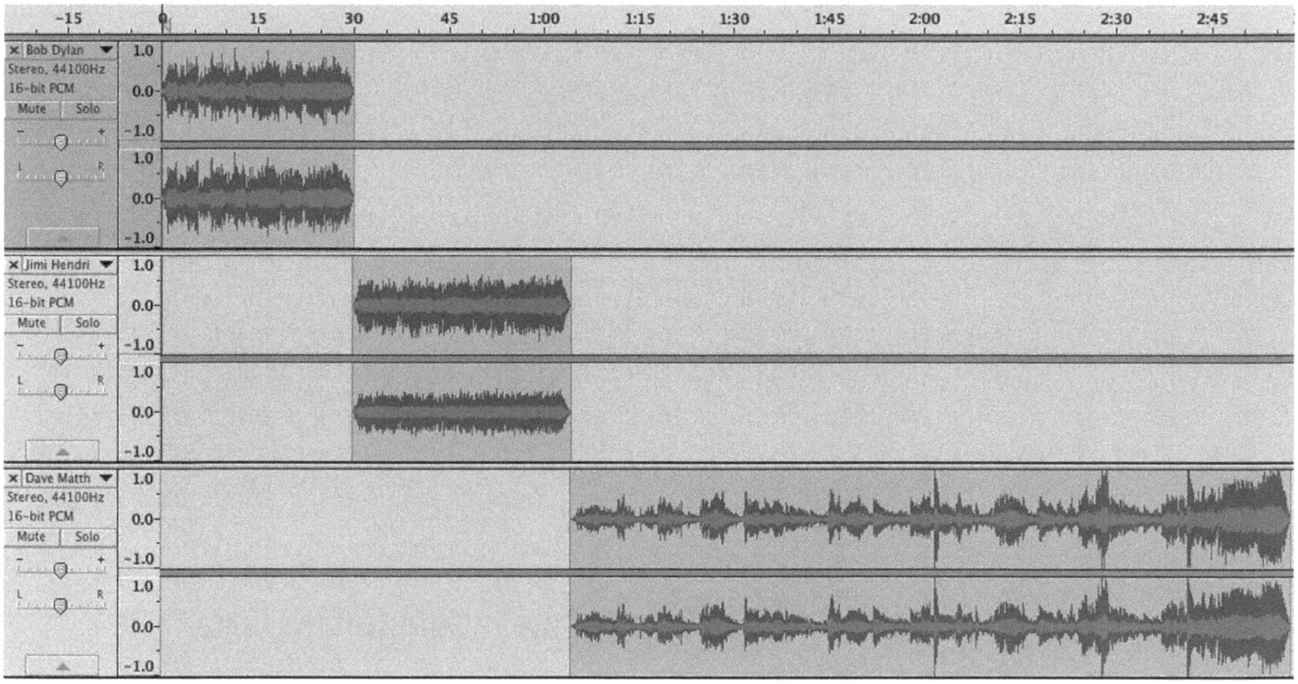

32. Export the completed file as an MP3.
    a. Select *File > Export Audio.*
    b. Set the *Format* to MP3.
    c. Name the file "All Along The Watchtower 3 Version" and save it to a memorable location.
    d. Fill in the appropriate information in the Metadata window.

TIP

As you start to create Audacity files with multiple tracks, try selecting *View > Fit Vertically* to adjust the size of your tracks so they are all visible vertically in your window.

Practice performing the melody along with the three artists in this Audacity file. Try to imitate the complete performance including any ornaments or variations to the the melody. The goal is to select your favorite ornaments/variations from each version to create your own interpretation of the melody.

TIP

Since these three versions are in different keys, you could choose to transpose the audio of two of them so they are all in the same key. Consider transposing Bob Dylan down a minor 2nd and Dave Matthews up a minor 3rd so all audio is in C minor (Jimi Hendrix's key). To do this, select Bob Dylan's audio and select *Effect > Change Pitch*. Set the *Semitones* value to *-1.00*. Select Dave Matthews's audio and select *Effect > Change Pitch*. Set the *Semitones* value to *3.00*.

NOTE: Yes, Dave Matthews does sound a little like Alvin and the Chipmunks with the pitch shift up of a minor 3rd.

TIP

As a variation to this activity, you may choose to select some of the instrumental "jam" sections between verses to compare/contrast the improvised solos that are also essential to this song. For example, compare Bob Dylan's harmonica solo to Jimi Hendrix's guitar solo.

Importing MIDI into GarageBand: "Maiden Voyage" by Herbie Hancock

[⊕] *See Example 4.3 on the companion website for a GarageBand file and audio recording example of the the "Maiden Voyage" activity.*

The software instrument sounds included in music production software are often superior to those found in accompaniment software. For this activity, we use Band-in-a-Box as a vehicle to quickly create an accompaniment and then use the exported MIDI file in GarageBand to improve the instrument sounds and add custom MIDI and audio recordings to the file.

"Maiden Voyage" by Herbie Hancock, on the album of the same name, presents a modal tune with a straight eighth-note feel in an AABA form. The A section uses A-7/D and C-7/F while the B section uses B♭-7/Eb and C♯-7. Below each chord are recommended scales (see Figure 4.13).

Figure 4.13

Chord progression and associated scales for "Maiden Voyage."

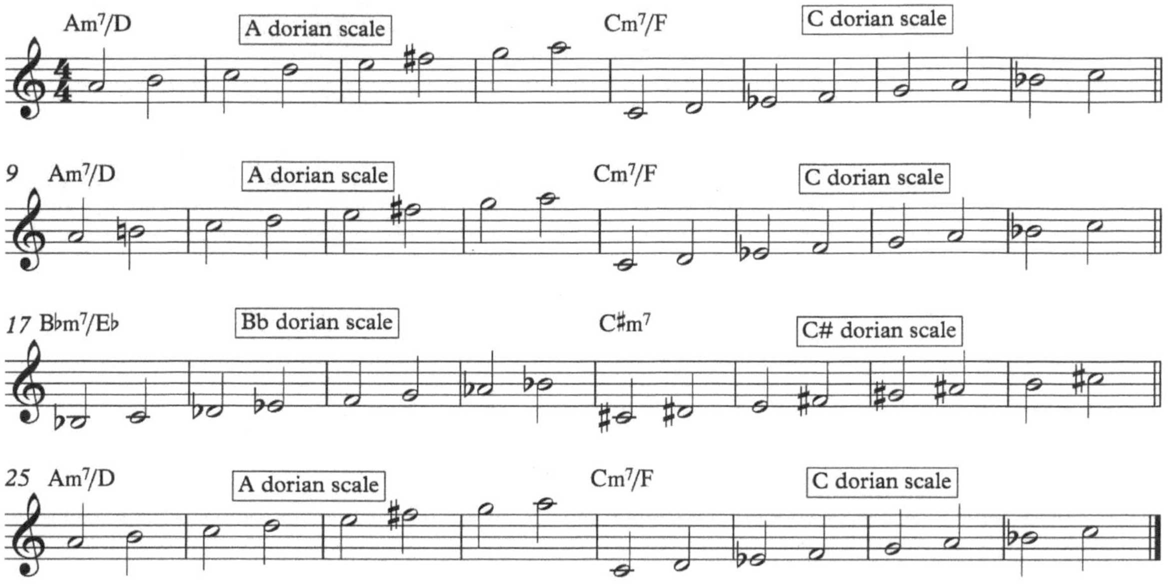

TIP

Explore the leadsheet for "Maiden Voyage" in Jamey Aebersold's *Volume 54: Maiden Voyage*. As previously mentioned, this play-along book includes many great tunes for beginning to intermediate improvisers including the title track. I use this book more than any other with my young improvisers.

Listen to and purchase "Maiden Voyage" on iTunes: https://itunes.apple.com/us/album/maiden-voyage-remastered/id721271378.

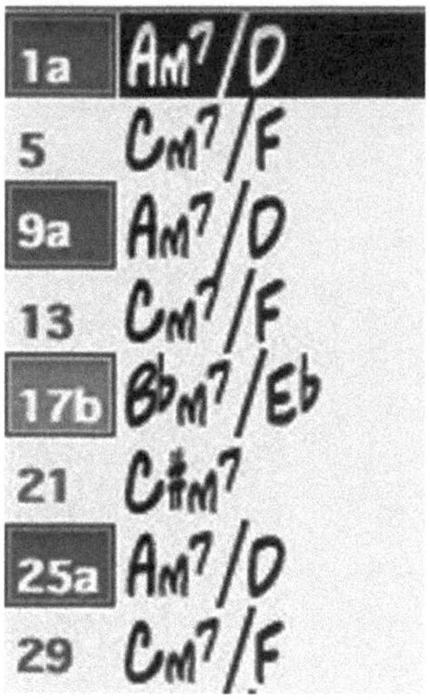

Figure 4.14

Chord progression for "Maiden Voyage" entered into Band-in-a-Box.

1. Launch Band-in-a-Box and enter the chords to "Maiden Voyage" listed above. Remember that the form is AABA (see Figure 4.14).
2. Select a style of your choice (preferably something in 4/4 with a straight-eighth feel). I selected *Modern Jazz #2* for my accompaniment.
3. Set the choruses to 4 so students can state the melody during chorus 1 and 4 and improvise during choruses 2 and 3.
4. Select *File > Save as MIDI*, select *File on Disk*, and save the MIDI file to a memorable location with the name "Maiden Voyage.mid."

Now that we have a MIDI file, we are ready to quit Band-in-a-Box and move to GarageBand.

TIP

If you are a PC user, consider purchasing Mixcraft. It is an excellent, affordable, user-friendly GarageBand equivalent. Mixcraft Home Studio 7 would be sufficient for completing all of the activities included in this text. Visit http://www.acoustica.com/mixcraft/ for more information.

5. Launch GarageBand. If an existing project opens, select *File* > *New* to open the startup window.
6. Select *New Project*, choose the *Empty Project* template, and select *Choose.*
7. Select *File* > *Save* to save your file to a memorable location.
8. Click and drag your Maiden Voyage.mid file into the GarageBand arrange window to import it.

TIP

You can easily change the software instrument for each track. Open the *Library* pane, select a track, and choose a new instrument sound. For example, I changed my bass sound from Muted Bass to Upright Studio Bass to achieve a more acoustic sound.

Now that the MIDI file is in GarageBand you have additional freedom to modify and edit the accompaniment to suit your needs. For this file, I suggest modifying the bass part during the melody choruses to more closely match Herbie Hancock's version.

9. Select the region on the bass track. Split the region (*Edit* > *Split*) at bars 3, 35, 67, 99, and 131.
10. Delete the bass part from 3–35 and 99–131.
11. With your MIDI keyboard connected (see Chapter 1 for more information on connecting a MIDI keyboard to your computer), record the bass accompaniment for the A section in bars 3–11. See Figure 4.15 for the A section bass accompaniment part.

Figure 4.15

Bass accompaniment for the A section of "Maiden Voyage."

12. Hold the OPT key on your QWERTY keyboard and click and drag to copy/paste this pattern to bars 11 and 27.

13. Click your playhead at bar 19 and record the bass accompaniment for the B section from bars 19–27. See Figure 4.16 for the B section bass accompaniment part.

Figure 4.16

Bass accompaniment for the B section of "Maiden Voyage."

14. Open the *Editor* pane by clicking the icon or press "E" on your typing keyboard.
15. Select the notes you recorded in the previous step and set the *Time Quantize* value to 8th note. This will help improve your rhythmic accuracy.
16. Hold the OPT key and click/drag to copy the material you recorded for bars 3–35 to bars 99–131.

To help guide students through the improvisation sections, we will create an audio recording announcing the upcoming scale. For this activity, I suggest using your built-in computer microphone to keep the process extremely simple. I discuss audio recording options in detail in the next activity.

17. Select *GarageBand* > *Preferences* and select the *Audio/MIDI* tab. Set your *Input Device* to *Built-in Microphone.*
18. Click the + icon in the upper left corner of the screen and select *Audio* > *Microphone* to create a new audio track.
19. Record your voice announcing each scale.
    a. A sections: A dorian and C dorian
    b. B section: B♭ dorian and C♯ dorian

TIP

I suggest announcing the scale 1 measure prior to when it hits. This allows the student to think ahead and prepare for the next key center. For example, announce "A dorian" in bar 34 in preparation for the arrival of A-7/D in bar 35.

TIP

You can adjust the input level of your built-in microphone if needed.

- Click the *Smart Controls* icon .
- Click the *Show Inspector* icon .
- Under *Record Settings*, adjust the *Record Level* up or down to improve your input level.

TIP

Be sure to use headphones when recording with a microphone to reduce the background sound.

20. Select *Share* > *Export Song to Disk.*
    a. Set the format to MP3.
    b. Name the file "Maiden Voyage Playalong."
    c. Click *Export* to save the MP3 to a memorable location.

See Figure 4.17 for a screenshot of a completed file.

Figure 4.17
Screenshot of completed "Maiden Voyage" GarageBand file.

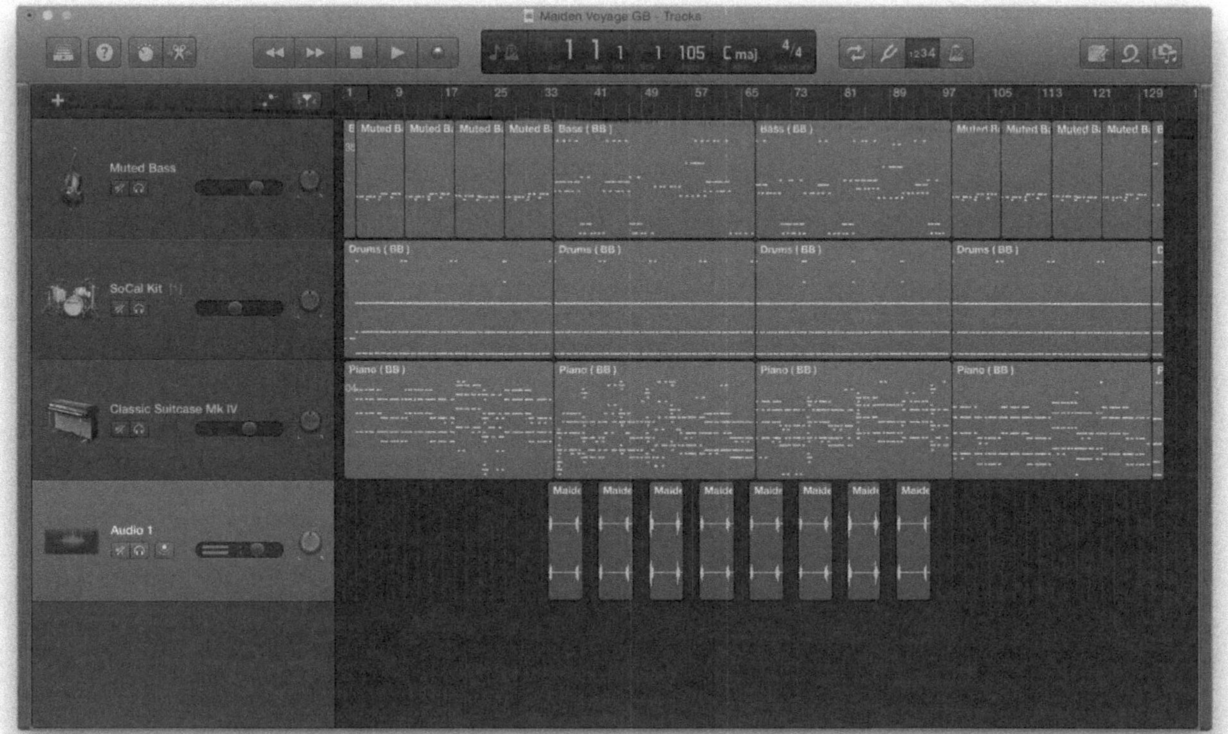

High-Quality Audio Recording in GarageBand and Audacity: Reprise of "All the Things You Are" Chord Tone Solo

[▶] *See Example 4.4 on the companion website for a GarageBand and audio recording example of the "All the Things You Are Chord Tone Solo" activity.*

For this activity we will reuse the chord-tone solo over "All the Things You Are" from Chapter 3. We will develop and export an accompaniment track in Band-in-a-Box. We will then focus on how to capture high-quality audio using both Audacity and GarageBand.

Prepare the Accompaniment File

1. Launch Band-in-a-Box.
2. Enter a title for the accompaniment and set the key to A♭.
3. Enter the chords for "All the Things You Are" (see Figure 4.18). See Chapter 3 for a discussion of the chord progression to this well-known tune.

Figure 4.18

Chord progression for "All the Things You Are" entered into Band-in-a-Box.

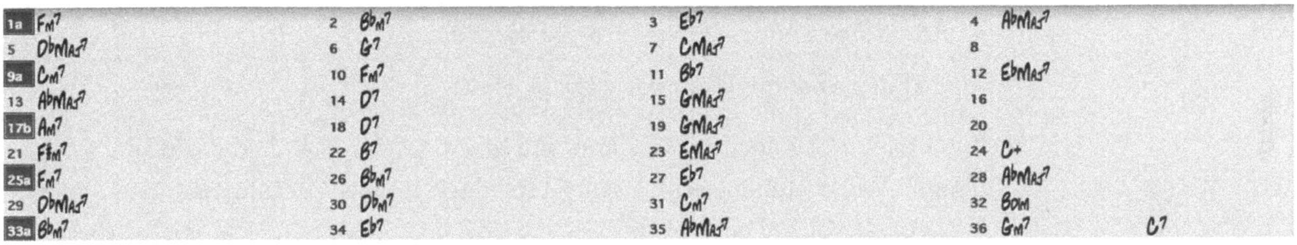

4. Set the *End* value to 36 and *Choruses* to 1 so the song only plays once through the form.
5. Select an appropriate style and adjust the tempo as needed.
6. Save your Band-in-a-Box file to a memorable location.
7. Select *File > Save Song as AIFF Audio* (Mac) or *WAV Audio* (Windows) and save the file to a memorable location. We will use this file for the Audacity portion of this activity.
8. Select *File > Save Song as MIDI File*. Choose *File on Disk* in the pop-up window and save this MIDI file to a memorable location. We will use this file for the GarageBand portion of this activity.

Microphone Options

Before actually doing a take, you must select appropriate equipment for recording. Over the past 10 years, the options for recording very high-quality audio have increased while the cost has decreased. Both Audacity and GarageBand can be used to recording top-notch audio; the quality is determined purely by your input equipment (microphone and audio interface). Below are my recommendations for capturing excellent quality audio recordings:

- Built-in Computer Microphone:

  When your budget is tight, using the built-in computer microphone is an option. The quality of built-in computer microphones is decent, and more important, there is no additional equipment to purchase. Be careful of playing directly into or too close to the built-in computer microphone as both can often cause distortion in these microphones really designed for video conferencing.

- USB Microphone:

  The Blue Snowball USB mic ($69.99) was one of the first professional-quality USB microphones available. Today, top manufacturers of traditional microphones including Rode, Audio Technica, and Shure are now in the USB microphone market and prices generally range from approximately $70 to $200. USB microphones have an A/D interface (A/D means Analog to Digital) built-in and connect directly to your computer with a USB cable—no additional gear required!

- Audio Interface with a Professional Microphone:

  Purchasing an audio interface allows you to connect any standard professional microphone to your computer. This setup provides the most flexibility for digital recording because you are not tied to a specific USB microphone and can freely experiment with various microphones to find the best sound. For recording a single instrument, you can purchase an audio interface with a single input such as the Focusrite Scarlett Solo ($99.99) or two inputs such as the M-Audio Fast Track II ($99.00) or Presonus AudioBox iTwo ($159.95). In terms of microphones, I suggest purchasing a well-known brand such as Shure, AKG, or Audio Technica. Off-brand microphones may have a lower price tag, but you typically get what you pay for. Visit a local music store and try out various microphone brands/models before making your investment. Personally, I use a Shure KSM27 (now discontinued, replaced with Shure SM27, $299.00) and have had success with this microphone on just about everything including saxophone, acoustic guitar, vocals, and large ensembles. Finally, you will also need some accessories including an XLR microphone cable (to connect the professional microphone to the computer audio interface), pop filter (for vocals only), and microphone stand.

Recording High-Quality Audio in AUDACITY

1. Launch Audacity.
2. Select *File* > *Save Project As* to save your new Audacity project to a memorable location.
3. Select *File* > *Import* > *Audio* and locate your exported Band-in-a-Box accompaniment **audio** file (WAV or AIFF) accompaniment.

TIP

If your audio levels appear quite low, improve them by selecting the audio and selecting *Effect* > *Normalize.*

4. Play the file. It should sound identical to how it sounded in Band-in-a-Box because it is a rendered audio file using the Band-in-a-Box instruments.
5. At the top of the Audacity screen you can set your various recording options. Set your audio input to your recording device (either Built-in Microphone, USB Microphone, or Audio Interface) and your audio output to your sound output device (Built-in Output or Audio Interface). Also set the channels to *1 (Mono) Recording Channel* since we are using a single microphone. See Figure 4.19 for my audio setup. I'm using the M-Track audio interface for input/output of audio along with a Shure KSM27 microphone.

Figure 4.19
Audio setup in Audacity for users working with an audio interface.

TIP

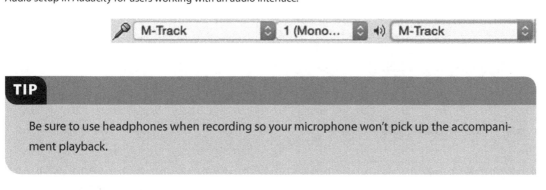

Be sure to use headphones when recording so your microphone won't pick up the accompaniment playback.

6. Click the *Pause* icon and then click the *Record* icon. This puts Audacity in stand-by record mode. Adjust the input levels in Audacity or the input level on your audio interface until the level meter indicates "good levels."

TIP

In the process of setting levels, we want the loudest levels to be close to but not above clipping. Clipping is generally displayed as RED in a level meter and clipping can cause audio distortion. Avoid clipping at all costs! I typically play my loudest volume on my instrument and adjust the levels up until I'm slightly below clipping. It is better to set levels slightly too low than slightly too high because we can use functions such as normalize to quickly improve soft levels.

7. Click the *Pause* icon again to begin recording. Perform the chord-tone solo you composed in Chapter 3 along with the accompaniment.
8. Select *Edit* > *Select All* to highlight all of your audio tracks.
9. Select *Effect* > *Reverb*. Adjust the settings to your liking. I suggest creating the style of reverb by adjusting the first 6 parameters (*Room Size* through *Tone High*) and then adjusting the balance of wet (reverb) and dry (direct) signal by reducing the *Wet Gain* until you have just enough reverb. You can preview the sound of

the reverb by clicking *Preview* and then apply the reverb to your audio tracks by clicking *OK*.

10. Adjust the volume balance between your tracks by adjusting the *Gain* slider for each track located below the track name. Be sure to keep an eye on your output level meter to avoid clipping!
11. Select *File > Export Audio*. Set the format to *MP3*, name the file, and save it to a memorable location.

See Figure 4.20 for a screenshot of the completed file.

Figure 4.20
Screenshot of completed "All the Things You Are" Audacity file.

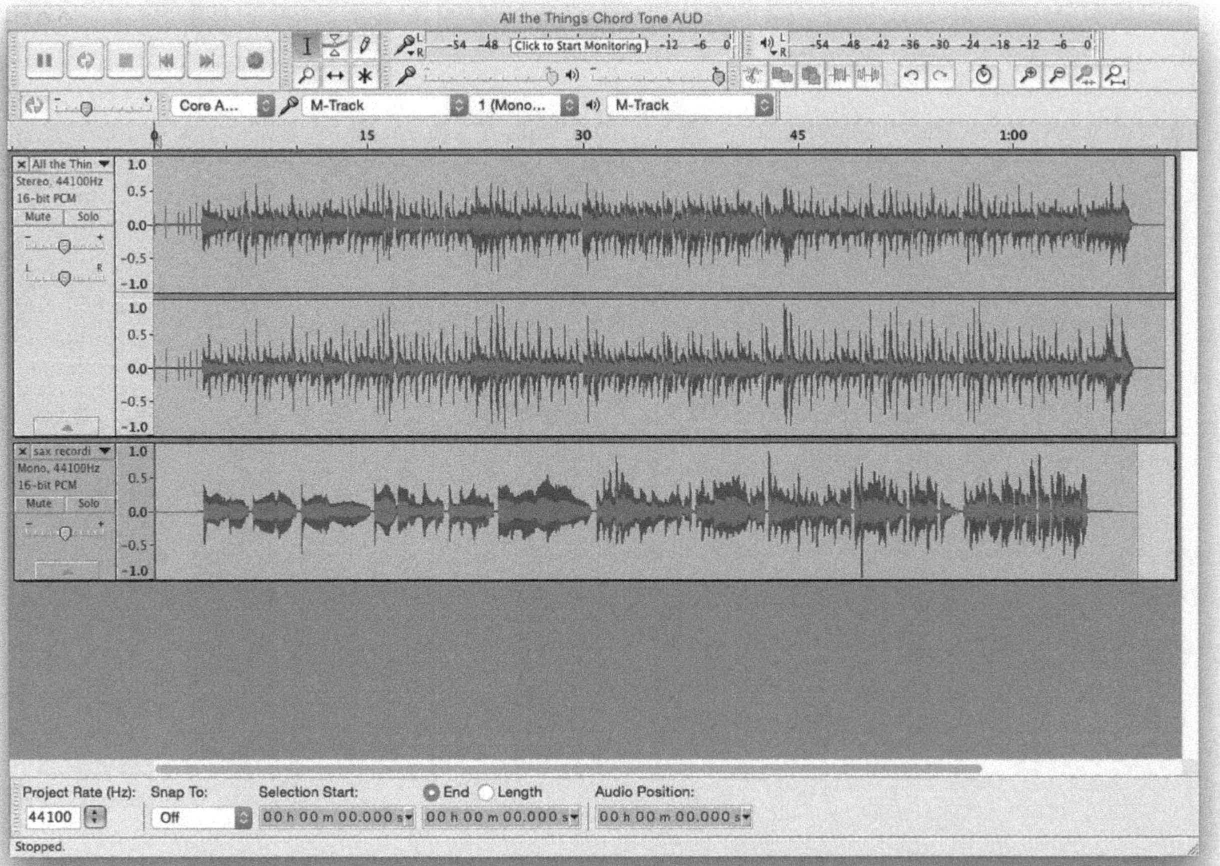

Recording High-Quality Audio in GarageBand

1. Launch GarageBand and create a new empty project.
2. In the pop-up *New Track* window, select the microphone icon. At the bottom of this window, select your audio input device (either Built-in Microphone, USB Microphone, or Audio Interface) and your audio output device (Built-in

Output or Audio Interface). Also set the channels to *Input 1* since we are using a single microphone.

TIP

In my studio setup with an M-Tracks audio interface and KSM27 microphone, I set "My instrument is connected with" to M-Track and "I hear sound from" to M-Track since I'm both recording and listening back through my audio interface.

3. Click the *Smart Controls* icon in the top left of the screen. In the *Smart Controls* window, click the *Show Inspector* icon. Click the triangle next to *Recording Settings* to expand this menu. Adjust the *Record Level* slider up and down to achieve good levels in the level meter in your audio track (see *Recording High Quality Audio in AUDACITY* for advice on setting levels).

TIP

If you are using a computer audio interface the *Record Level* slider will likely be grayed out because you adjust your input level on the interface box itself.

4. Save the file to a memorable location.
5. At this point you have two choices: import the Band-in-a-Box audio file OR the MIDI file. Since GarageBand has fairly nice sounding software instruments, let's import the MIDI file. Click and drag the file into your GarageBand file.
6. GarageBand will automatically select the appropriate instruments. Feel free to experiment with various software instruments sounds in the Library for each track.
7. The default tempo is 120 bpm. Adjust the tempo as needed in the top center *before* recording any audio!
8. Click on the empty audio track you created in step 2.
9. Click the *Record* icon at the top of the screen to begin recording and click the *Stop* icon to end the recording.

TIP

You can turn the count-in or metronome click on/off by clicking on these icons at the top of the screen.

TIP

Click the *Input Monitoring* icon to hear yourself as you are recording. Adjust the volume slider on the track to balance the level of the accompaniment and your live performance.

NOTE: The *Input Monitoring* icon is available in the *Smart Controls* menu for audio and guitar tracks. Be sure to click the *Show Inspector* icon and click the triangle next to *Recording Settings* to expand this menu. Avoid feedback by using headphones while monitoring during recording.

10. Return to *Smart Controls* and click *Show Inspector* . Scroll down and click the triangle next to *Plug-ins* to expand this menu. Add reverb to your recording by adjusting the *Master Reverb* Master Reverb: ✓ slider up until you have just enough reverb.

TIP

Consider clicking on an empty plug-in slot and selecting one of the available reverb plug-ins. This allows more control of the style of reverb you add to your recording.

11. Adjust the panning of each track by turning the *Pan* dial . I suggest keeping the audio recording and drums panned center while the other accompaniment instruments (bass, piano, guitar, etc.) are panned varying degrees left and right.
12. Select *Track* > *Show Master* from the menus at the top of the screen to open the master track in your arrange window.
13. Click the Master Track to select it.
14. Click the *Library* icon in the upper left corner. Select one of the mastering presets available in the window. These mastering presets add a variety of effects including EQ (to adjust the overall tone of the file) and compressor/limiter (to squash the loudest sounds and make the entire file feel louder). Listen to the sonic differences as you select the various mastering presets until you find one that enhances the overall sound of your file.

TIP

Open the *Smart Controls* window and click *Show Inspector* and *Master* Master to open the master effects from any track. You will now see each plug-in added from the preset you selected in the library. You can turn these plug-ins on/off by clicking the *Power* icon , adjust the plug-in setting by clicking the *Settings* icon , or select a different effect for that slot by clicking the *Up/Down Arrow* icon .

NOTE: You must hover over the plug-in to make these icons display.

TIP

Most mastering presets include a limiter plug-in at the end of the signal chain. Click the *Settings* icon next to the limiter plug-in and slowly adjust the *Gain* setting up to increase the overall volume of the audio output.

NOTE: You decrease the dynamic range of the file by doing this so be judicious with how much you increase the *Gain* and let your ears guide you.

15. Select *Share* > *Export Song to Disk*. Set the format to *MP3*, name the file, and save it to a memorable location.

See Figure 4.21 for a screenshot of the completed file.

Figure 4.21
Screenshot of completed "All the Things You Are" GarageBand file.

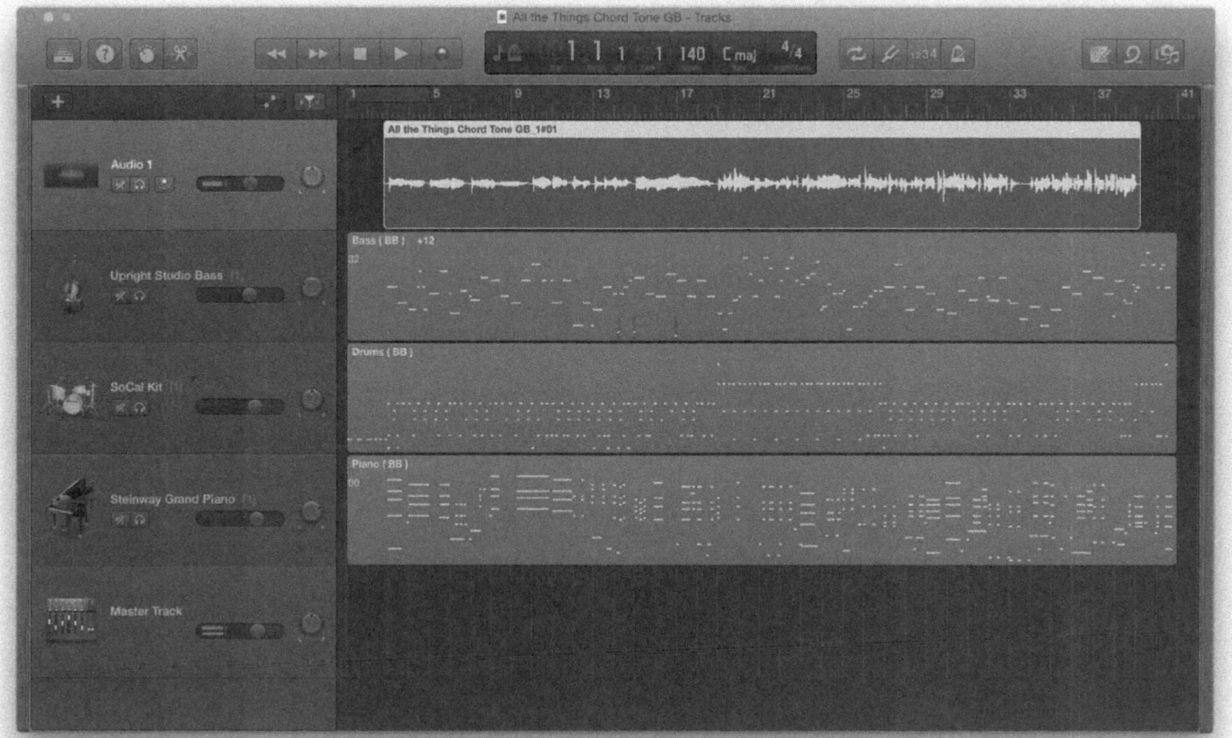

iPad Connection: Creating Accompaniments Using Smart Instruments in GarageBand for iOS (GBiOS): "Little Sunflower" by Freddie Hubbard

[▶] *See Example 4.5 on the companion website for an audio recording example of the the "Little Sunflower" activity.*

Composed and performed by trumpet great, Freddie Hubbard, "Little Sunflower" is a beautiful modal AABBAA tune that combines minor and major tonalities with a straight-eighth groove. You can find a lead sheet for this tune in *Volume 60: Freddie Hubbard* by Jamey Aebersold.

Examine the simple form to this tune in Figure 4.22.

Figure 4.22

Chord progression for "Little Sunflower."

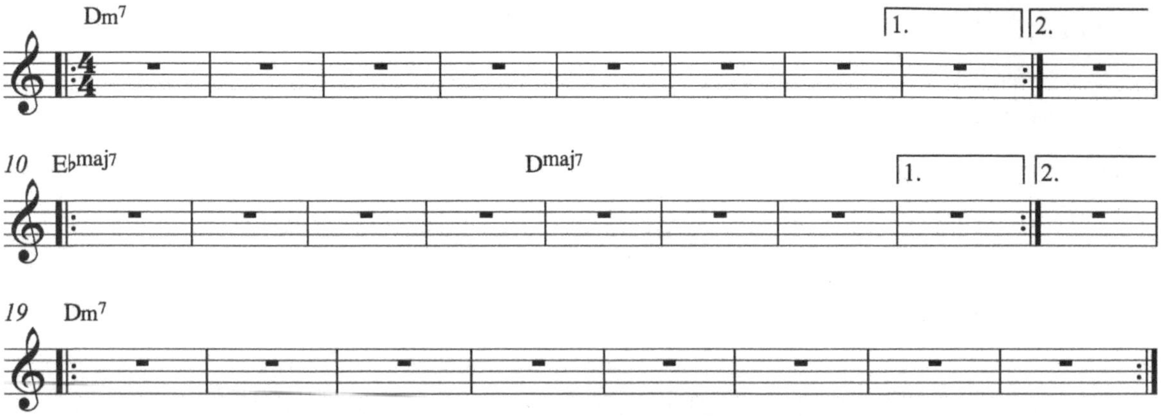

Knowing just three scales (D dorian, E♭ major, and D major), students can improvise on "Little Sunflower" (see Figure 4.23). It is also relatively easy for young improvisers to *feel* and *hear* the change in chords since the change from D minor to E♭ major is so obvious and the change from E♭ major to D major occurs at 4 bar intervals.

Figure 4.23

Scales used in "Little Sunflower."

Listen to and purchase "Little Sunflower" by Freddie Hubbard on iTunes: https://itunes.apple.com/us/album/little-sunflower-lp-version/id76147210?i=76147149.

For this activity, we will explore GarageBand for iOS (GBiOS) as the vehicle to create our own custom accompaniment track for "Little Sunflower." Through the use of Smart Instruments, GBiOS provides the opportunity to quickly create a unique play-along. We will also record an improvised solo using the scale restriction available in the GBiOS Keyboard instrument.

For more information on GBiOS, visit http://www.apple.com/ios/garageband/.

To download GBiOS, visit https://itunes.apple.com/us/app/garageband/id408709785?mt=8.

1. Launch GBiOS and create a new project by tapping the "+" icon in the upper left corner and select *Create New Song*. NOTE: If your GBiOS opens into an existing file, you will need to first tap "My Songs" My Songs in the upper left corner.
2. Swipe to the right and select *Drummer. Drummer* is a new virtual instrument that mimics a real drummer playing various beats and fills.
3. Tap the name of the drummer (see Figure 4.24). Scroll through the other drummers available in GBiOS. Each drummer has a brief description of his or her style. Select a drummer you'd like to use for your play-along file.
4. Scroll through the drum presets next to the drummer's name and select one that you like for your A section of "Little Sunflower" (see Figure 4.24). Each preset will change the style of the beat.

Figure 4.24 GBiOS *Drummer* selection screen. Choose your drummer and a preset.

TIP

You can customize the drum pattern by modifying the various settings at the bottom of the screen.

1. *XY Grid*: Drag the yellow dot around the square to change the pattern from simple to complex and soft to loud.
2. *Swing Menu*: Set the swing amount. For this song, I suggest selecting *Swing None*.
3. *Kit Pieces*: Tap the various instruments on the drum image to set the instrumentation of the pattern.
4. *Fills*: Modify the intensity of the drum fills at the end of the phrase.
5. *Patterns*: Select a different pattern from the three sliders on the right.

Now that the drum beat is set for the A section, we are ready to add in chords and bass using Smart Instruments.

5. Tap the *Tracks View* button.
6. Tap the "+" icon in the lower left to add a new instrument.
7. Select the *Smart Guitar*.

TIP

Tap the guitar picture to select a new guitar sound. You can choose between Acoustic, Classic Clean, Hard Rock, and Roots Rock.

8. Tap the *Song Settings* button.
    a. Tap *Tempo* and adjust the tempo of the song. For this song I suggest 130 bpm.
    b. Tap *Key* and set the key to D minor.
    c. Tap *Edit Chords*. Set the first chord to D-Min-7, second chord to D♯/E♭-Maj-Maj7, and the third chord to D-Maj-Maj7 (see Figure 4.25). Tap *Done* to exit this window.

Figure 4.25 Dm7, E♭Maj7, and DMaj7 chords programmed into the *Smart Guitar* in GBiOS.

TIP

These chords will carry over to the other Smart Instruments in this song!

9. Tap the *Record* button and perform a chordal accompaniment for the A section using only Dm7.

TIP

Select one of the autoplay patterns and just tap the chord once.

10. Tap the *Tracks View* button.
11. Tap the "+" icon in the lower left to add a new instrument
12. Select the *Smart Bass*.
13. Tap the *Record* button and perform a bass accompaniment for the A section using only Dm7.

TIP

Similar to the Smart Guitar, tap the bass image to select a new bass sound. I prefer *Upright* for this song.

TIP

Improve the rhythmic accuracy of your recording by tapping the *Mixer* button in the upper right corner and setting the *Quantization* value. You will set the quantization to the smallest rhythmic value of your performance. For faster rhythms try 16th note, slower rhythms try 8th note, and very slow rhythms try quarter note.

NOTE: I avoid quantizing strummed guitar tracks because the quantization eliminates the rolled chord sound.

14. Tap the "+" in the upper right corner to open the *Song Sections* menu.
15. Tap *Duplicate* to create a second copy of your first section to create the "AA" part of the "AABBAA" form of this tune.
16. Tap *Add* to create a new section. GBiOS is now calling this *Section C* but it really will serve as the B (bridge) of "Little Sunflower."
17. Double-tap the Section C to open it.
18. Double-tap the *Drummer* track.
19. Modify the beat to create a contrasting bridge section.
20. Once you are satisfied with your contrasting drum beat, tap the *Tracks View* button .
21. Double-tap the *Smart Guitar* track.
22. Tap the *Record* button and perform a chordal accompaniment for the bridge section using E♭Maj^7 for 4 bars and DMaj^7 for 4 bars.
23. Tap the *Tracks View* button .
24. Double-tap the *Smart Bass* track.
25. Tap the *Record* button and perform a bass accompaniment for the bridge section using E♭Maj7 for 4 bars and DMaj7 for 4 bars.
26. Tap the *Tracks View* button .
27. Tap the "+" in the upper right corner to open the *Song Sections* menu.
28. Tap *Duplicate* to create a second copy of your bridge to create the "BB" part of the "AABBAA" form of this tune.
29. Tap the *Section A* and then tap *Duplicate* twice to create the final "AA" part of the tune.
30. Tap *All Sections* to view your entire play-along file (see Figure 4.26).

Figure 4.26

Accompaniment for AABBAA form of "Little Sunflower" performed into GBiOS.

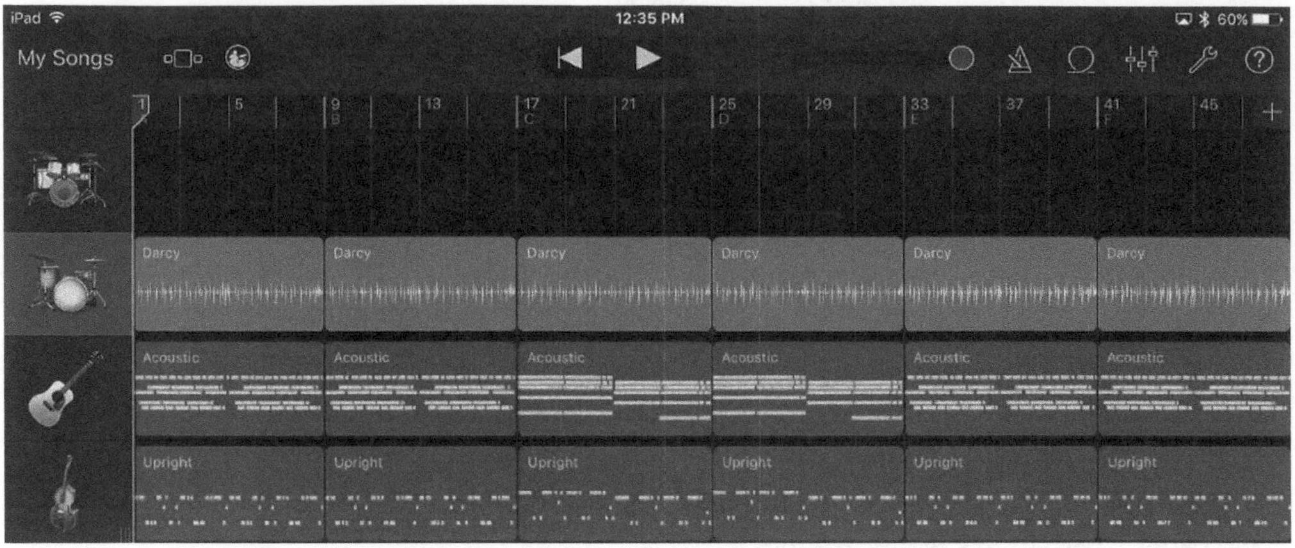

31. Tap the "+" icon in the lower left to add a new instrument and select the *Keyboard* (NOT *Smart Keyboard*).
32. Tap the *Scale* icon in the Keyboard instrument window and set the scale to *Dorian*. You have now restricted your on-screen keyboard to only the notes from the D dorian scale.
33. Tap the *Record* button and perform an improvised solo over the first AA (bars 1–16) and last AA (bars 33–end) of the tune.
34. Tap the *Scale* icon in the keyboard and set the scale to *Major* to restrict the notes to only the notes from the D major scale.
35. Tap the *Record* button and perform an improvised solo over the last four bars of each bridge section (bars 21–24 and 29–32).
36. Tap the *Song Settings* icon and tap *Key*. Turn OFF *Follow Song Key* and select E♭ major (see Figure 4.27). Tap anywhere on the screen to exit this menu.
37. Tap the *Scale* icon in the keyboard and set the scale to *Major* to restrict the notes to only the notes from the E♭ major scale.
38. Tap the *Record* button and perform an improvised solo over the first four bars of each bridge section (bars 17–20 and 25–28).

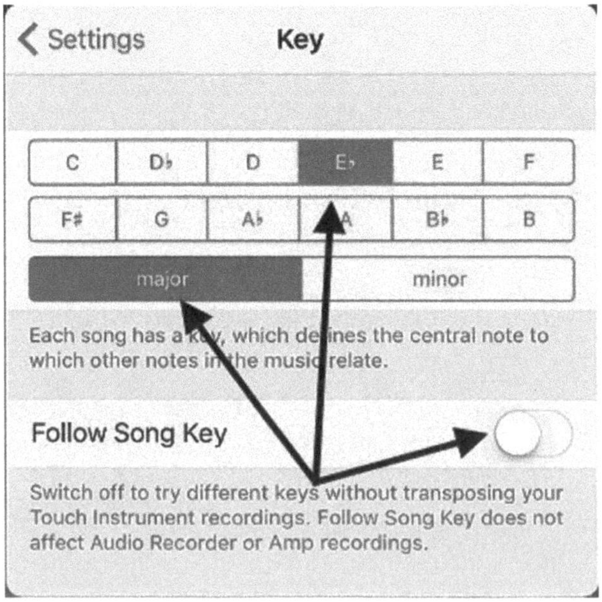

Figure 4.27

Set the song key to E♭ Major and turn OFF *Follow Song Key*.

39. Tap *My Songs* My Songs in the upper left corner to save and exit your file.

TIP

I suggested exiting a song every once in awhile to allow GBiOS to save your file since there is no other *Save* option.

40. Tap the name plate of the song you just created. Name the song "Little Sunflower Play-along" and tap *Done.*
41. Tap and hold on your play-along file. The files will jiggle and there will be a blue border around your file.
42. Tap the *Share* icon and select *iTunes*.
43. In the *Send to iTunes* window, select *iTunes* (not *GarageBand*) to export your completed audio file.
44. In the *Share Song* window enter the appropriate information under *My Info,* select the desired format (typically *Medium Quality* or *High Quality*), and tap *Share.*
45. Connect your iPad to your computer via the USB cable and open iTunes.
46. Click on your iPad in the devices area (see Figure 4.28).
47. Click on *Apps* in the left pane and scroll down to the *File Sharing* section (see Figure 4.28).
48. Click on *GarageBand* in the *Apps* area below *File Sharing* (see Figure 4.28).
49. Locate "Little Sunflower Play-along.m4a" and drag the file to your desktop (see Figure 4.28). You now have the completed file available to add to your iTunes library, email to students, post to a class website, etc.

Figure 4.28

Connect your iPad to your computer via the USB cable and open iTunes. Click on your iPad in the devices area. Click on *Apps* in the left pane and scroll down to the *File Sharing* section. Click on *GarageBand* in the *Apps* area below *File Sharing*. Locate "Little Sunflower Play-along.m4a" and drag the file to your desktop.

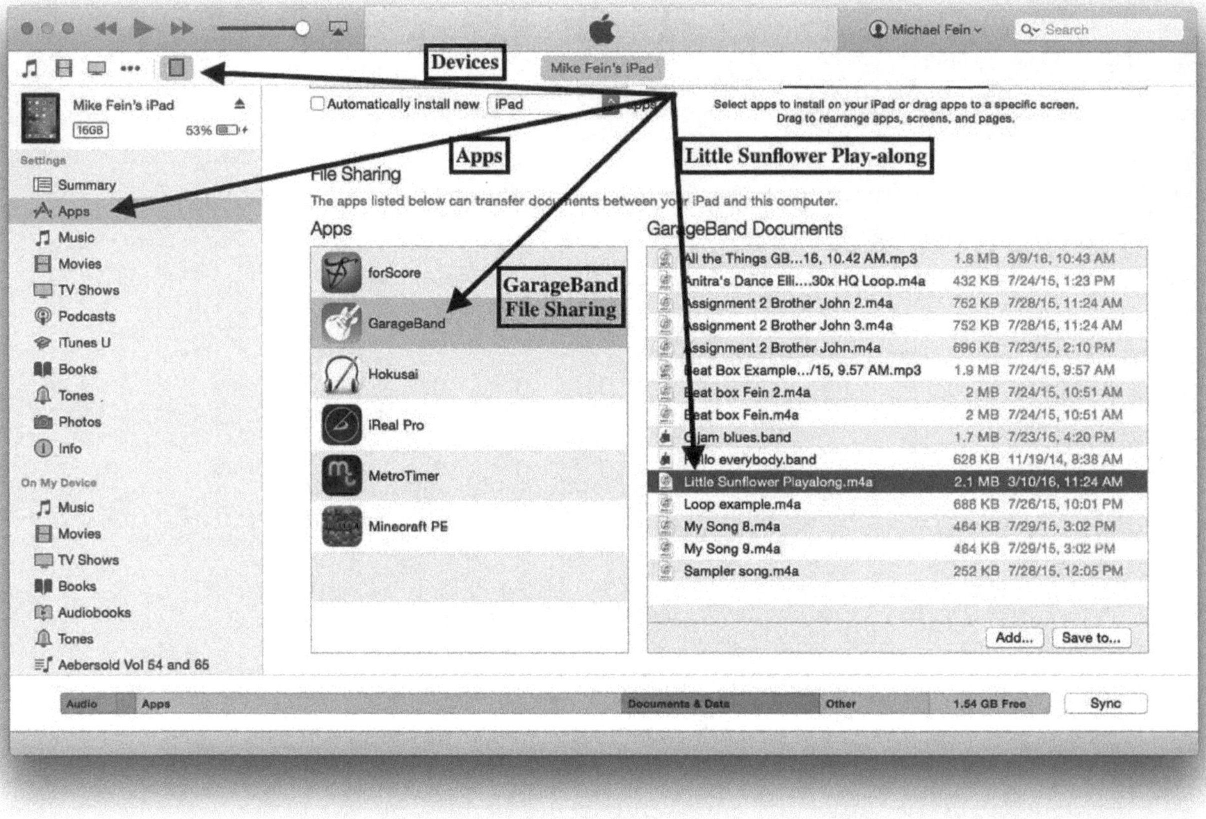

Chapter 4 Review

1. Using either your own transcription or one you located on the Internet, analyze Miles Davis's use of the dorian mode throughout his "So What" solo. NOTE: I located a set of transcriptions by Mauro Guenza at http://etpmusic.com/archives/208.
2. Using Audacity, import "While My Guitar Gently Weeps" from *The White Album* by the Beatles. Isolate the iconic Eric Clapton guitar solo (approx. 2:00–2:30). Add a smooth fade in/out, remove all other audio material, slow the tempo, and export the file as MP3. Transcribe the performance using notation software or search the Internet for a transcription and practice along with the MP3.
    a. Listen and purchase this song on iTunes: https://itunes.apple.com/us/album/while-my-guitar-gently-weeps/id401126224?i=401126346.
    b. Explore the story behind Clapton's guest appearance on this recording from the following *Guitar World* blog post:
        i. http://www.guitarworld.com/features-blogs/hear-eric-claptons-isolated-guitar-track-beatles-while-my-guitar-gently-weeps/%0914575).

3. Compare and contrast the recordings of "All Along the Watchtower" as performed by Bob Dylan, Jimi Hendrix, and Dave Matthews. Be sure to use musical terms in your response such as tempo, instrumentation, melodic ornamentation, key signature, etc.
4. What are the pros and cons of using a USB microphone versus a professional microphone with a computer audio interface?
5. In this chapter, we used Audacity, GarageBand, and GBiOS to manipulate audio/MIDI and create accompaniment files. Create a chart comparing the pros and cons of each application as they relate to teaching improvisation.

WEB RESOURCES FOR LISTENING 5

YouTube and Spotify

It has never been so easy to access media of great improvisers in action using the Internet. My two favorite places to locate and organize media are YouTube and Spotify. YouTube has an incredible amount of content (both audio and video) of live performances and commercial recordings and users can create playlists to gather their favorite videos into a single location. I have also provided links for some of my favorite YouTube channels that include performance recordings and instructional videos. Of course, some of the media posted to YouTube is in clear violation of copyright and is always in danger of being taken down at anytime. Spotify is completely legal alternative that includes the ability to create playlists of your favorite tracks. This service contains only commercial audio recordings (no video and no bootlegs or unofficial material) and it is completely free for a basic account.

Using YouTube to Teach Musical Improvisation

Although you can view YouTube content without creating a YouTube account, you must have a YouTube account to create custom playlists, so I strongly urge you to create an account. Playlists will allow you to compile a list of videos with a couple of clicks when you are signed in to your YouTube account.

TIP

If you have a Google account (Gmail, Google Drive, etc.) you already have a YouTube account.

Here are the steps to create a FREE YouTube account:

1. Visit www.youtube.com.
2. Click *Sign In* Sign in in the upper right corner.
   a. If you already have a Google account, sign in on the next screen and skip ahead to step 3 below.
   b. If you do not have a Google account:
      i. Click *Create Account.*
      ii. Enter the requested info.

> **TIP**
>
> When creating a new Google account, you can use an existing email address OR create a new Gmail email account.

Next, you will create your own YouTube channel. Think of a YouTube channel as the homepage for your YouTube account. Your channel will include any videos you upload as well as custom playlists that you create (more on creating playlists in a moment).

3. Click your profile picture in the upper right corner of the screen and click the *Gear* icon to access *YouTube Settings*.
4. You are now in the *Overview* screen. Under *Additional Features*, click *See all my channels or create a new channel* (see Figure 5.1).

Figure 5.1

In the *Overview* screen, under *Additional Features*, click *See all my channels or create a new channel.*

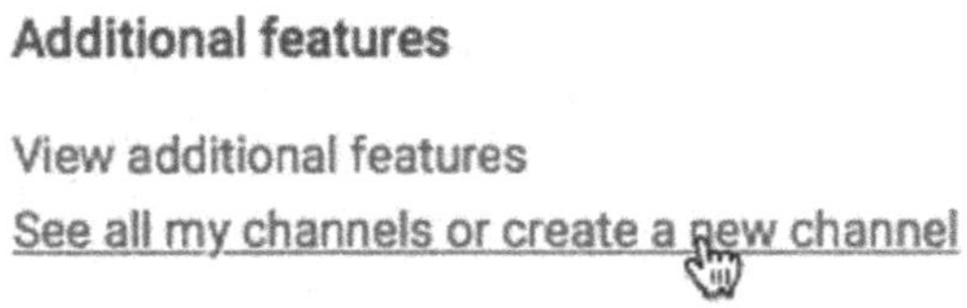

5. Click *Create a new channel* .
6. Name your new channel, select a category, check the box for *I agree to the Page Terms*, and click *Done*. You may need to verify your account before proceeding.
7. Customize your new channel by adding channel art and a channel description.

You are now ready to browse videos and create a playlist.

[▶] *See Example 5.1 on the companion website for a link to an example YouTube playlist for this activity.*

8. Enter a query into the *Search* field at the top of the screen.
9. Click any video in your search to open it.
10. Below the video, click the *Add to* button + Add to.
11. Click *Create New Playlist* Create new playlist. Name the new playlist and click *Create*.

> **TIP**
>
> As you upload content or create playlists, you will have the option to set the privacy settings to public, unlisted, or private.
>
> - Public—Anyone can search for and view your content.
> - Unlisted—Your content will not show up in a search but anyone can view your content.
> - Private—Your content is restricted to only those you specifically allow.

TIP *continued*

Generally for education, I suggest setting your content to *Unlisted*. This allows you slightly more control over who is viewing your material than *Public*; only students with the specific link will be able to access your content and not just the general YouTube user (since your content will not be searchable). I almost never use *Private* since this content is so restrictive.

12. Navigate back to your search list and open a different video.
13. Again, below the video, click the *Add to* button + Add to. This time, you will see the playlist you created in the previous step. To add the new video to the playlist, simply click the checkbox next to the playlist name. ✓ Teaching Musical Improvisati...
14. Continue finding videos and adding them to your playlist.
15. Click the *Guide* icon in the upper left of the screen and click on *My Channel* My Channel. You should now see your new playlist in the *Created Playlists* section.

TIP

You can modify the order of the videos in your playlist.

- In the *My Channel* screen, click *Created Playlists*.
- Click the name plate of your playlist to open the playlist editor.
- Hover your mouse over a video until you see a gray vertical bar near the video's thumbnail. Click and drag the video up and down to reorder the playlist.

16. Hover your mouse over the thumbnail image for your playlist and you should see the words *Play All*. Click the thumbnail and your first video will begin playing and you will see a listing of the other videos in the playlist on the right side (see Figure 5.2).

Figure 5.2

When you click on a playlist, the first video plays on the left side while a listing of the other videos in playlist displays on the right side.

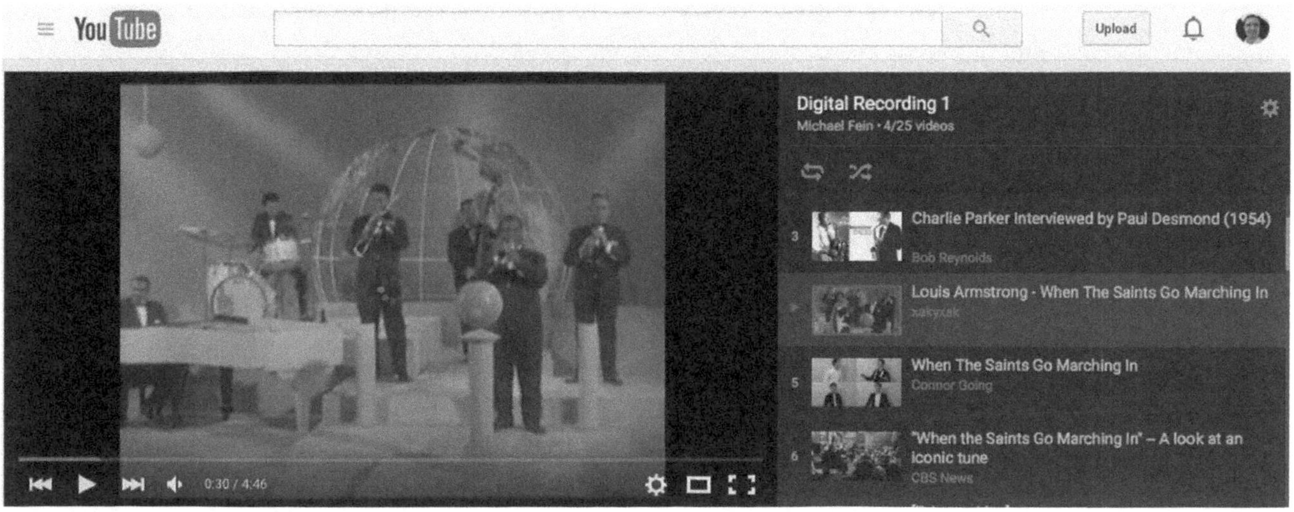

17. Click the *Share* icon Share under your video. You can share this playlist via link, embed, or email. For now, the best option is to use the link. Highlight and copy the link and paste it into an email or existing website (see Figure 5.3).

Figure 5.3 Click the *Share* icon and copy/paste the link to share a link to your video/playlist.

> **TIP**
>
> In Chapter 7 of this text, I explain how to embed and/or post YouTube videos and playlist links to your own website. Embedding is my favorite way to share YouTube content!

> **TIP**
>
> Create a playlist that includes multiple versions of the same song. This gives students a chance to hear a variety of improvisational approaches to the same chord changes.

Recommended YouTube Channels for Improvisation Material

The amount of material available on YouTube is vast but you can whittle it down to a manageable amount by finding and subscribing to outstanding YouTube channels. Similarly, my cable TV package has a few hundred channels, but I know I generally enjoy programming on the History Channel, NBC, ESPN, etc., and I usually check those stations first for my evening entertainment.

Here are some of my go-to YouTube channels for improvisational material:

- Jazz at Lincoln Center's JAZZ ACADEMY—https://www.youtube.com/user/jalcjazzacademy—Led by musician/educator Wynton Marsalis, Jazz at Lincoln Center has one of the best music education channels on YouTube. This channel features lessons on a variety of jazz topics by jazz greats and live performances of exceptional young musicians.
- Denis DiBlasio—https://www.youtube.com/user/1DDBBBB—Denis DiBlasio is the director of Jazz Studies and Composition at Rowan University in Glassboro, New Jersey. I had the pleasure of studying with Denis for three years while I attended Rowan University to earn my master's degree in Jazz Performance. Denis is an

incredible jazz saxophonist/flautist with a gift for working with beginning to advanced musicians.

- Learn Jazz Standards—https://www.youtube.com/user/Learnjazzstandards—This channel includes play-alongs for jazz standards, recordings of jazz greats, articles on jazz, special features such as *Lick of the Week*, and more.
- Jeff Schneider—https://www.youtube.com/user/JSchneidsMusic—I stumbled on Jeff's channel one day and I have really enjoyed viewing his music lessons.
- Andreas Eastman—https://www.youtube.com/user/AndreasEastmanWinds/featured—This instrument manufacturer's channel has lots of great videos from outstanding jazz eductors such as Chris Farr and Tony White.
- Blue Note Records—https://www.youtube.com/user/BlueNote—This is the official channel of the iconic record label with recordings and interviews of Blue Note artists.
- Solo Transcriptions with On-screen Notation—As previously discussed in this book, transcribing and analyzing improvised solos will have a huge impact on your musical development. Explore these channels that feature transcribed solos with notation and synchronized audio.
  - Carles Margarit—https://www.youtube.com/channel/UCL5GDSOXfQ8QhOlvg-Gpkisg
  - Mauro Guenza—https://www.youtube.com/channel/UClsJXI4U5hlZfnCNlj8hKGA

Subscribe to your favorite channels to find them more quickly in the future. Subscribing also lets the channel's author know that people are paying attention to his or her videos and encourages the author to make more great videos.

1. Login to your YouTube account and visit one of the channels above.
2. Click the red *Subscribe* button Subscribe next to the author's name.
3. Click the *Guide* icon YouTube in the upper left corner and then click *Subscriptions* Subscriptions. This page shows the latest uploads from your subscribed channels.
4. Click the *Guide* icon YouTube again and scroll down to *Manage Subscriptions* Manage subscriptions. Here you can unsubscribe from channels by hovering over the *Subscribed* button and clicking *Unsubscribe* Unsubscribe.

TIP

You can also unsubscribe from a channel directly from the channel home page or while viewing a video from that channel. Simply click *Unsubscribe* Unsubscribe next to the author's name.

The YouTube Copyright Dilemma

Copyright is a major issue for YouTube users, so much so that YouTube has an entire section of its website devoted to copyright education at https://www.youtube.com/yt/copyright/.

Just about every piece of media is covered by copyright law. As soon as an audiovisual work (movie, TV show, other online video), sound recording, visual work, or written work is in any tangible form it is protected by copyright law. This means that no one else can use the work without permission from the copyright holder—PERIOD. Posting copyright protected material is an infringement even if you credit the copyright owner, post a disclaimer, and make zero financial gains from the video. You are even violating copyright if you record yourself performing a copyright protected work. So all singers/songwriters who record and post themselves performing non-public domain cover songs are violating copyright unless they were granted permission from the copyright owner.

TIP

Public domain works include those published before 1923, between 1923 and 1963 (but the copyright was not renewed), or before 1977 (but did not contain an appropriate copyright notice). Since use of works in the public domain is not restricted by copyright law, anyone is free to record and post a cover version of a public domain song. Visit PD Info (http://www.pdinfo.com/public-domain-music-list.php) for a searchable list of public domain music.

The only loophole for using non-public domain material is the fair use doctrine. Whether something falls under fair use is ultimately decided by a judge based on four factors, and all four factors must be met to qualify under fair use. The four factors of fair use are as follows:

1. *The Purpose and Character of the Use*: Commercial uses are generally not fair while educational or non-profit uses generally are fair.
2. *The Nature of the Copyrighted Work*: Using material from fictional or creative works is generally not fair while using material from informational sources generally is fair.
3. *The Amount and Substantiality of the Portion Used in Relation to the Copyrighted Work as a Whole*: Using a small portion (10% or 30 seconds, whichever is less) is generally permissible.
4. *The Effect of the Use on the Potential Market for, or Value of, the Copyrighted Work*: Uses that limit the copyright holder's ability to profit from his or her original work are generally not fair.

If all of this starts to get confusing, you're not alone. This is why lawyers get paid the big bucks! The most import thing you can do is educate yourself about copyright law and try your best to follow the guidelines as you understand them. If you ever post something to the Internet and receive a takedown notice or some other official correspondence, remove your questionable content immediately.

I delve deeper into copyright as it pertains to posting media to the Internet in Chapter 6, Web Resources for Posting, but for now the main concern is how this impacts playlists and video links. Since YouTube is a public venue, any video guilty of copyright infringement is subject to removal. YouTube videos that violate copyright are removed either because of a content ID match discovered by YouTube software that scans videos for copyrighted material in a large database or a takedown notice submitted by the copyright holder claiming that a video has violated copyright. It is extremely common to find a great video, often of a live performance or commercial recording, only to discover that it has been removed the very next day for copyright reasons. So, as you create playlists of great YouTube material, avoid videos that you suspect violate copyright as their life on YouTube will often be short-lived and you will find deleted videos on your beloved playlist!

Using Spotify to Teach Musical Improvisation

Spotify is a music streaming service launched in 2008 and it has completely changed the way I consume music. For the purposes of teaching musical improvisation, Spotify is a great source for locating examples of outstanding improvisers and sharing this audio with students. Unlike YouTube, all of the music you enjoy on Spotify is 100% legal and there are never any copyright issues because artists are paid per stream according to current royalty rates. Users can stream music and create/share playlists from the desktop application or from a mobile device all for FREE! Spotify does have a paid Premium package that provides certain advantages including no ads, offline music, and improved sound quality. Additionally, on mobile devices (except iPad/tablet) free users can only shuffle playlists while premium users can select specific songs for playback. For more information about Spotify Premium visit https://www.spotify.com/my-en/premium/.

TIP

If you teach in a classroom with access to a desktop or laptop computer or tablet with an Internet connection, I suggest sticking with the free version of Spotify. If your teaching situation requires you to stream content from your mobile device, then the paid premium account is worth the monthly investment.

To get started with Spotify, you must make an account. This is one hurdle to overcome with Spotify because, unlike YouTube, users cannot start listening until they have created an account. This means that all of your students need to sign up with Spotify to listen to one of your shared playlists.

1. Visit www.spotify.com and click on *Sign Up*.
2. Sign up with your Facebook account or your email address.
3. After creating your account, you need to download the application. This may download automatically but if it doesn't, visit https://www.spotify.com/my-en/download/ to download the Spotify application.

You are now ready to organize material for easy access. The first method is to use playlists.

[▶] *See Example 5.2 on the companion website for a link to an example Spotify playlist for this activity.*

4. Launch the application and sign in.
5. Scroll down on the left pane and click *New Playlist* 

.

6. Name the playlist.
7. Search for songs by artist, album, or song title in the search field in the upper left corner Search.
8. Hover over one of the songs in the search list. Click the *More* icon for more options, select *Add to Playlist* Add to Playlist, and choose the playlist you created in step 5.
9. Continue to add songs to this playlist.

You can also organize music in your library by saving the song/album or following the artist.

Figure 5.4
Click the artist name in the search results.

10. Perform a search for an artist such as "Miles Davis."
11. Click the artist name in the search results (see Figure 5.4).
12. You are now in the artist view screen. Click *Follow* FOLLOW to save this artist.
13. Scroll down and click the + icon next to any song to save the song. The icon will change to a check mark after you save the song (see Figure 5.5).

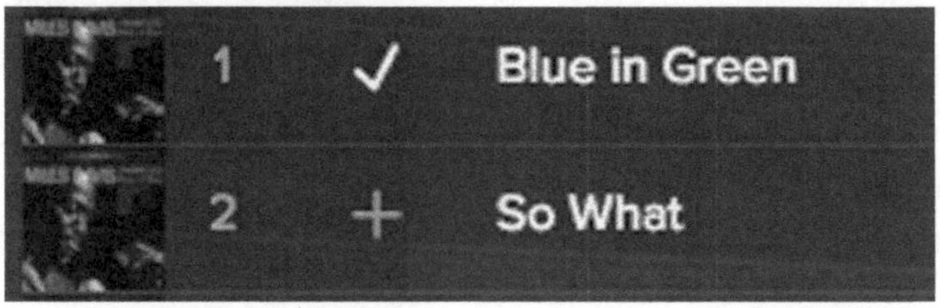

Figure 5.5
Click the + icon next to any song to save the song. The icon will change to a check mark after you save the song.

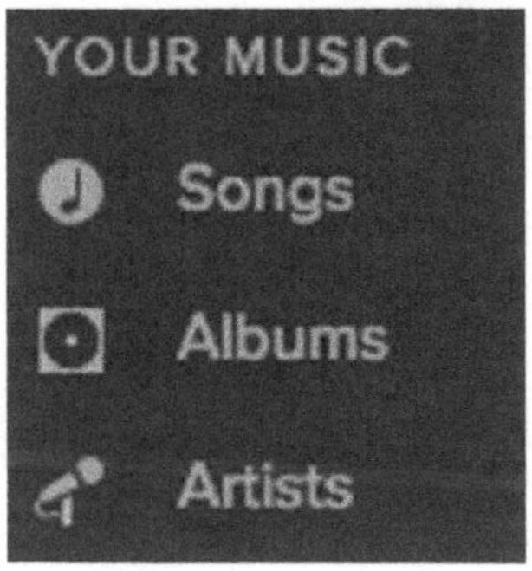

Figure 5.6
Below *Your Music*, click on each item (*Songs*, *Albums*, and *Artists*) to explore your saved content.

14. Click *Save* SAVE next to one of the albums such as *Kind of Blue*.
15. On the left pane of the screen, locate *Your Music* (see Figure 5.6). All of the items you saved/followed in the previous steps now exist in this section. Click on each item (*Songs*, *Albums*, and *Artists*) to explore your saved content.

To share saved songs:

16. Below *Your Music* click *Songs* (see Figure 5.6).

17. Click the *More* icon next to any song.
18. Select *Copy Song Link.*
19. Paste this link into an email or website for student access. When students click on the link, the Spotify application will launch and begin playing the song.
    a. NOTE: Students must have a Spotify account and application to listen to the audio material.

To share saved albums:

20. Below *Your Music* click *Albums* (see Figure 5.6).
21. Click one of the albums to open *Album* view.
22. Click the *More* icon below the album name.
23. Select *Copy Album Link.*
24. Paste this link into an email or website for student access.

To share saved playlists:

25. Click the playlist in the left pane.
26. Click the *More* icon below the playlist name.
27. Select *Copy Playlist Link.*
28. Paste this link into an email or website for student access.

TIP

You cannot share an artist that you are following.

TIP

In Chapter 7 of this text, I explain how to embed and/or post Spotify songs, albums, and playlists to your own website. Embedding is my favorite way to share Spotify content!

iPad Connection: YouTube and Spotify Apps

The YouTube and Spotify iPad app experience is very similar to the desktop experience. Since all of your saved videos/playlists/etc. are tied to your account, you should notice that everything ports directly to your iPad app once you login.

Note: Hold the iPad in landscape mode throughout this chapter.

Download the YouTube iPad app:
https://itunes.apple.com/us/app/youtube/id544007664?mt=8.

Download the Spotify iPad app:
https://itunes.apple.com/us/app/spotify-music/id324684580?mt=8.

Using the YouTube iPad App

1. Launch the YouTube app on your iPad.
2. Tap the *Account* icon in the top row.
3. Tap *Sign In* .
4. If you already have Google apps installed on your iPad, you may notice your Google accounts listed. Select one of these existing accounts or tap *Add Account* and enter your credentials.
5. Tap the *Home* icon to view your home page. Swipe up to see new and recommended videos or tap the *Search* icon and enter a new search.
6. Tap any video to open it.
7. While viewing the video:
   a. Tap *Subscribe* in the upper right corner next to the author's name to subscribe to his or her channel (see Figure 5.7).
   b. Tap the *Playlist* icon at the top right corner of the video (see Figure 5.7). Select *Add to Playlist* and choose an existing playlist or create a new one.
   c. Tap the *Share* icon at the top right corner of the video (see Figure 5.7). Select *Email* to email a link to this video or select *Copy Link* to copy/paste a link to this video into any document/app.

TIP

If you don't see the *Playlist* icon, tap the video once.

Figure 5.7
While viewing a video in the YouTube iPad app, subscribe to the channel, add the video to a playlist, or share a link to the video.

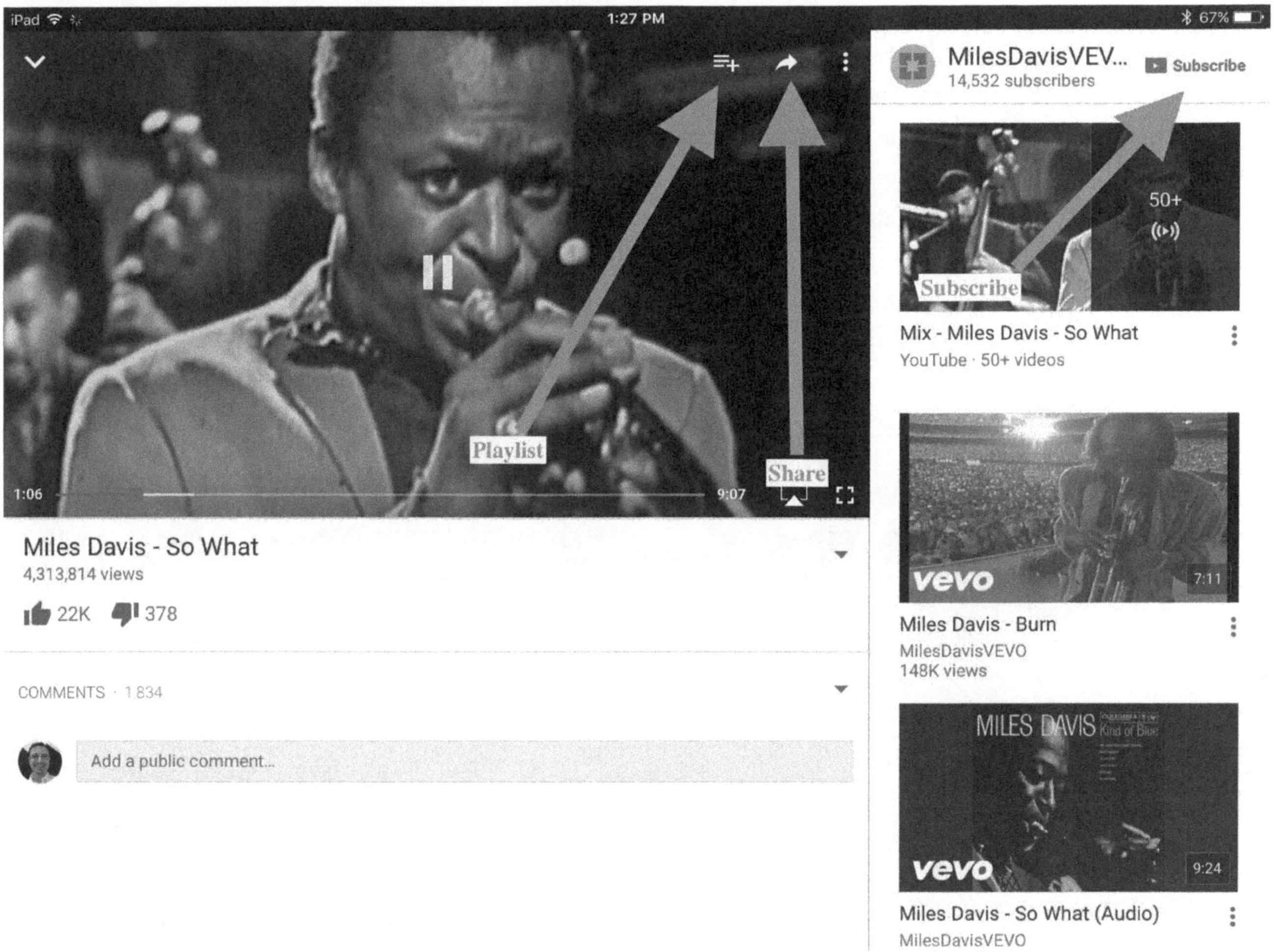

8. Close the video when you have finished.

TIP

To minimize a video, swipe the video down to the lower right corner. Close the video by swiping the minimized version to the right. My 3-year-old-son Grayson taught me that trick!

9. Tap the *Subscriptions* icon in the top row. All of your subscribed channels should now display.
10. Tap any channel in the left pane and then any video on the right to begin watching a video. Close the video when you have finished.
11. Tap the *Account* icon in the top row.
12. Swipe up to view all of your playlists.
13. Tap one of your playlists to open it.
    a. Tap the red *Play* arrow to begin playback of your playlist (see Figure 5.8). As the first video plays, tap the name of the playlist in the upper right corner to view the other videos in the playlist.
    b. Tap the *Share* icon to email or copy the link to your playlist (see Figure 5.8).
    c. Tap the *More* icon and select *Edit* to rename the playlist, add a description, or change the privacy settings (see Figure 5.8).

TIP

You cannot currently change the order of the playlist on the iPad app. You must login to your account on your Mac/PC computer to do this.

Figure 5.8

Open one of your custom playlists. Tap the *Play* icon to begin playback. Tap *Share* to share a link to your playlist. Tap *More* to edit your playlist.

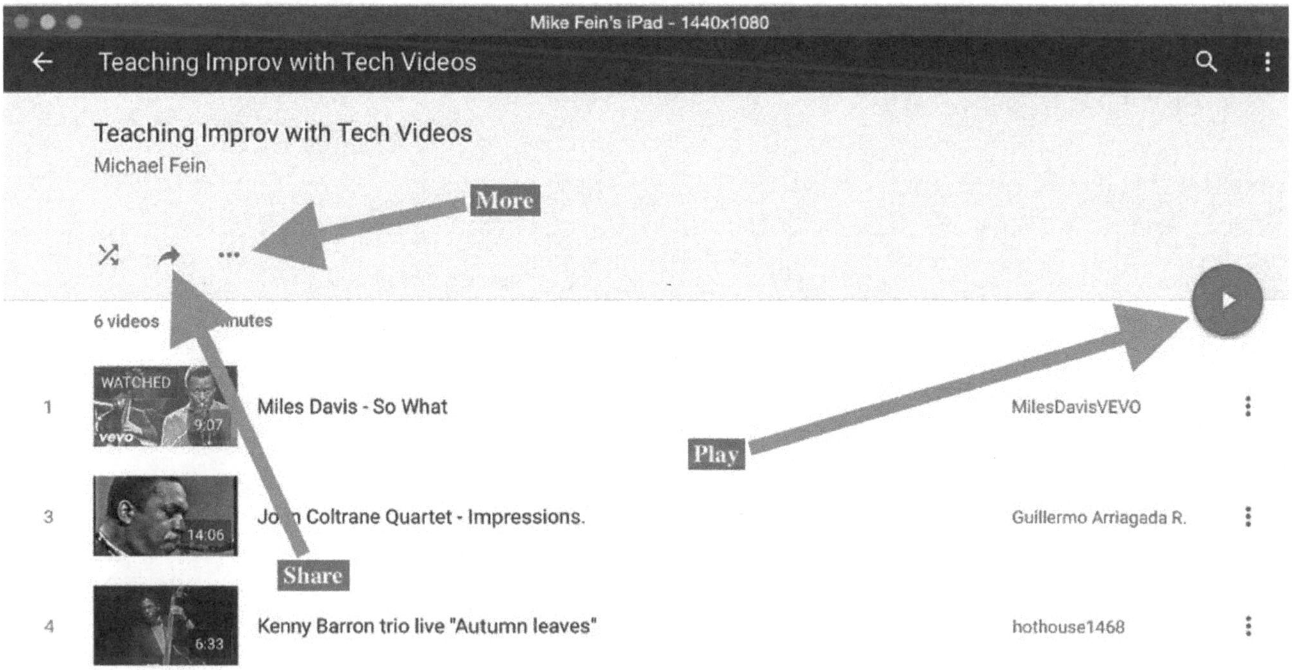

Using the Spotify iPad App

1. Launch the Spotify app on your iPad.
2. Tap *Log In* and enter your credentials.
3. Tap *Search* in the left pane to find new content.
4. In the search results:
   a. Tap one of the artists. Tap *Follow* to save this artist to your library.
   b. Tap one of the albums. Tap *Save* to save this artist to your library.
   c. Tap one of the songs. The song begins playing.
      i. Tap the + icon to save this song to your library. The icon will turn into a *Check* icon.
      ii. Tap the *More* icon and select *Add to Playlist*. Tap the + icon to create a new playlist or select an existing playlist.
      iii. Tap the *Minimize* icon to to minimize the song and view the full interface.

TIP

For free accounts, shuffle play is your only playback option on non-iPad/tablet devices such as an iPhone.

5. Tap *Your Music* in the left pane.
6. Tap *Playlists, Songs, Albums,* or *Artists* to view your saved material.
   a. Tap a song to begin playback.
7. Tap the *More* icon next to any playlist, song, or album and select *Share*.
   a. Select *Send to . . .* to send the link via email.
   b. Select *Copy Song Link* to copy a link to this playlist/song/album that you can paste into any document/app on your iPad.

Chapter 5 Review

1. What are the benefits and drawbacks of using YouTube for listening and instructional material?
2. Discuss the copyright law as it relates to YouTube. Specifically focus on the four factors of fair use and how/why videos are removed from YouTube for copyright infringement.
3. What are YouTube channels and why is it helpful to subscribe to channels?
4. What are the benefits and drawbacks of using Spotify for listening material?
5. What are the benefits of using playlists on YouTube and Spotify and how can you share playlists with students from these services?

WEB RESOURCES FOR POSTING 6

SoundCloud, Podomatic, Google Drive, and YouTube

Throughout this book, you have created a wealth of material and now we will explore how to post this material to the web for students to access. In the past, teachers could share material with students by dubbing a tape or burning a CD. Today, all of this can be accomplished both in a public or private venue without transferring any physical media. SoundCloud, Podomatic, and YouTube are three services that allow users to post material to the web, mostly in the public venue. Google Drive is another, more private venue for sharing files. This cloud storage option allows users to share files of any kind or even entire folders with students, and since the sharing is private, there are fewer copyright concerns.

In this chapter, I focus purely on uploading a file to each of the services mentioned above. In Chapter 7, I explain how to use a website to organize all of this material in one central location.

Preparing Audio Files for Web Posting

Export Audio Review

Let's look back at the various applications we have worked with in Chapters 2 through 4 of this book and review how we can export audio files appropriate for web posting.

- Band-in-a-Box: Select *File* > *Save Song as .m4a audio*
- iReal Pro: Share > Share Audio > AAC.
- Note Flight: Select *Action* > *Score* > *Export*. Choose *WAV audio*.
- Audacity: Select *File* > *Export Audio* and set the format to *MP3*.
- GarageBand: Select *Share* > *Export Song to Disk* and set the format to *MP3* or *AAC*.
- GarageBand for iOS: From the *My Songs* screen, tap and hold on a song until it jiggles. Tap the *Share* icon and select *iTunes*. In the *Send to iTunes* window select *iTunes*. In the *Share Song* window enter the appropriate information under *My Info*, select the desired format (typically *Medium Quality* or *High Quality*), and tap *Share*. Connect your iPad to your computer via the USB cable, open iTunes, and click on your iPad in the devices area. Click on *Apps* in the left pane of iTunes. Scroll down to the *File Sharing* section and click the *GarageBand* app. Locate your exported file and drag it to your desktop.

Converting Audio File Using iTunes

iTunes is a great application for listening to, purchasing, and organizing music on your Mac/PC but it is also a great audio tool capable of converting audio files to more Internet-friendly formats. In this case, I cover the steps for converting a WAV file exported from Note Flight into an MP3 using iTunes.

1. Launch iTunes on your Mac/PC.
2. Select *iTunes* > *Preferences* (Mac) or *File* > *Preferences* (PC).
3. Click the *General* tab at the top (see Figure 6.1).
4. Click *Import Settings* (see Figure 6.1) and set the format to *MP3*. Click *OK* twice to exit the preferences menu.

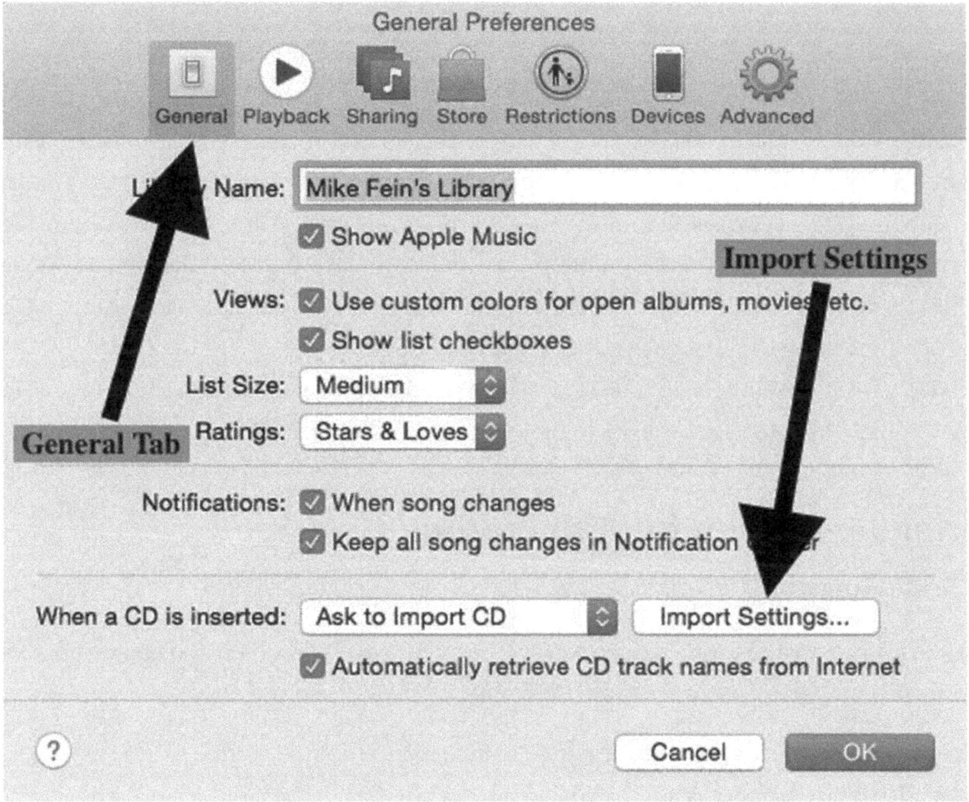

Figure 6.1
Click the *General* tab at the top and click *Import Settings*.

5. Drag your exported Note Flight WAV file into iTunes to import it.
6. Click the file to select it.
7. Select *File* > *Create New Version* > *Create MP3 Version*.

You now have an MP3 file that is about one-tenth the size of the previous WAV file.

TIP

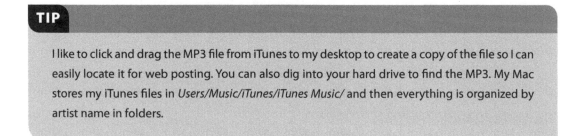

I like to click and drag the MP3 file from iTunes to my desktop to create a copy of the file so I can easily locate it for web posting. You can also dig into your hard drive to find the MP3. My Mac stores my iTunes files in *Users/Music/iTunes/iTunes Music/* and then everything is organized by artist name in folders.

TIP

Since WAV files are about 10 times the size of comparable quality MP3 or AAC files, I'd suggest only uploading MP3 or AAC files to SoundCloud or Podomatic.

Posting to SoundCloud

[▶] *See Example 6.1 on the companion website for a link to my SoundCloud account page.*

SoundCloud is a web-based audio service that allows users to post music and other audio files. Other users can view and even comment on your audio material. I choose to post audio to SoundCloud because most of my students are already listening to music on SoundCloud and they are comfortable with the interface. I have also found the service very easy to use.

As with most services you must first create an account.

1. Visit www.soundcloud.com and click on *Create an Account* Create account.
2. Enter your email address and create a password or connect your new SoundCloud account to your FaceBook or Google Plus account.

TIP

When you created a YouTube channel, you also created a Google Plus account. Feel free to link this account to SoundCloud.

You can upload any WAV, MP3, or AAC file that you have created in the various activities in this book. As previously mentioned, stick with MP3 or AAC files when possible due to the file size issue of WAV files.

TIP

Be sure you are not violating copyright law before uploading your files to a public forum like SoundCloud. Based on my understanding of copyright law, you are protected by the fair use doctrine by posting any of the files we created in this book as long as you omit the melody (chord progressions are not copyrightable but melodies are) and use 10% or 30 seconds or less of copyrighted material.

3. Click *Upload* Upload at the top of the site.
4. Click *Choose a File to Upload* Choose a file to upload and locate the audio file you want to upload on your computer hard drive.
5. Enter a title, genre, tags, and a description.

TIP

Tags help others locate your audio file through searches. The more descriptive the tags, the better the chance that someone will find your audio and listen. Type a word or phrase and hit the *RETURN* key to generate the tag with the hashtag prefix. An example tag maybe "jazz playalong" and this phrase would display as Additional tags #jazz playalong.

TIP

You have the option of setting your audio upload to *Private*, but since sharing audio with the world is the purpose of web posting, I'd stick with *Public*.

6. Click *Home* Home at the top of the site. Your latest upload should be at the top of your *Stream* Stream in this window.
7. Click the *Share* icon below any of your uploaded audio files (see Figure 6.2).

Figure 6.2
Click the *Share* icon below any of your uploaded audio files.

8. Copy the link and paste as needed into an email or webpage.

TIP

Encourage students to *Follow* your SoundCloud account so they can easily locate your posted audio files. To locate your account:

- Invite students to search "Your_Account_Name" in the *Search* box at the top of the SoundCloud site.
- Click *People* People on the left side and then click the *Follow* icon Follow below the account name.

You can also provide a link to your main page:

TIP *continued*

- Click your account name on any of your posted material (see Figure 6.3).
- Copy/paste the website URL from your web browser (see Figure 6.3).

Figure 6.3

To provide a link to your SoundCloud page, click your account name on any of your posted material, copy the website URL from your web browser, and paste it into an email/website.

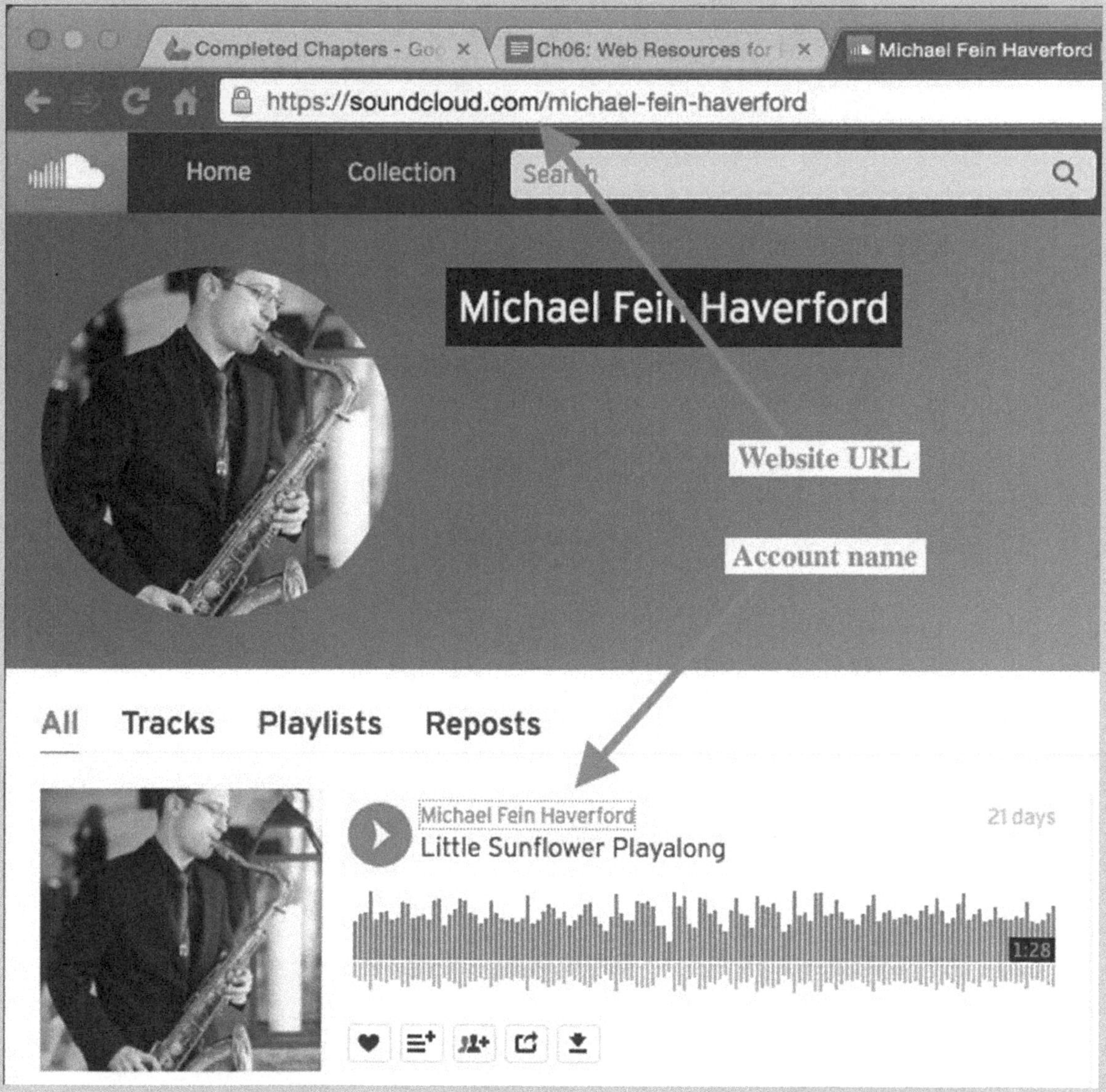

TIP

As users listen to your audio file, they can post comments at specific spots in your song.

To post a comment to someone else's audio file:

- Be sure you are logged into your SoundCloud account.

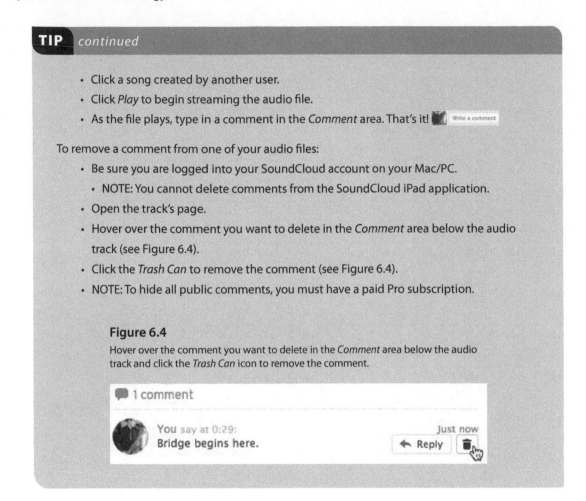

TIP *continued*

- Click a song created by another user.
- Click *Play* to begin streaming the audio file.
- As the file plays, type in a comment in the *Comment* area. That's it! 

To remove a comment from one of your audio files:

- Be sure you are logged into your SoundCloud account on your Mac/PC.
  - NOTE: You cannot delete comments from the SoundCloud iPad application.
- Open the track's page.
- Hover over the comment you want to delete in the *Comment* area below the audio track (see Figure 6.4).
- Click the *Trash Can* to remove the comment (see Figure 6.4).
- NOTE: To hide all public comments, you must have a paid Pro subscription.

Figure 6.4

Hover over the comment you want to delete in the *Comment* area below the audio track and click the *Trash Can* icon to remove the comment.

Posting to Podomatic

[▶] *See Example 6.2 on the companion website for a link to my Podomatic account page.*

Podomatic is a free podcasting service. There are two primary benefits to podcasting:

1. Submit your podcast to iTunes to make your content available on the world's largest music store for FREE.
2. Students can subscribe to your podcast once and continually get new episodes downloaded to their iTunes account automatically.

TIP

There are literally thousands of free podcasts available on the iTunes store from every topic imaginable. Try searching for podcasts on iTunes and subscribing to those that interest you. Once you start exploring the world of podcasts you will be hooked!

The first time you use Podomatic you need to create an account, setup your podcast with a podcast URL, and submit the URL to iTunes. Each time you post after that, the process goes even quicker and your material is live on Podomatic.com as well as iTunes.

NOTE: Initial navigation of the Podomatic site can be slightly confusing. I've included direct web links through these steps to help you locate the various sections of Podomatic.

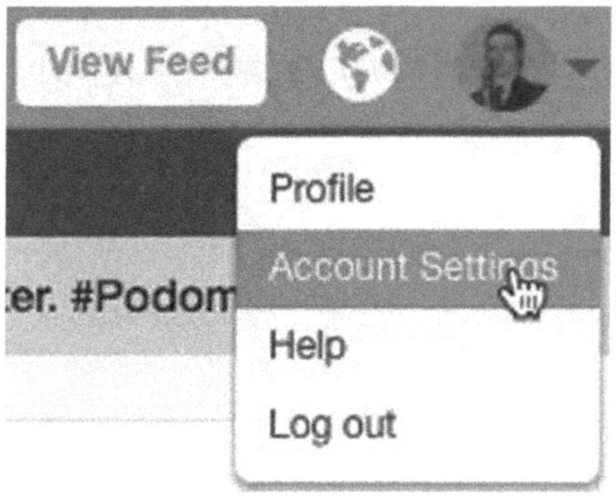

Figure 6.5
Click your avatar in the top right corner and select *Account Settings*.

1. Visit www.podomatic.com and click *Sign Up* Sign up.
2. You can choose *Sign Up with Email* or *Sign Up with Facebook*. Enter your credentials to create your Podomatic account.
3. Click your avatar in the top right corner and select *Account Settings* (see Figure 6.5).
4. Under *Settings Area*, in the upper right click *Podcast Settings* (http://www.podomatic.com/settings/podcast).
5. Under the *Basic Info* tab, click *Choose a Podcast URL* and enter a URL for your podcast.

TIP

Be careful when selecting your podcast URL because you can't go back and change it later!

6. Navigate to your podcast management page (http://www.podomatic.com/podcast).
7. Click the *Publish* menu Publish at the top of the screen and select *Upload* Upload from the dropdown list. (http://uploads.podomatic.com/mymedia/uploader).
8. Import some digital photos/images using one of the methods (I suggest the HTML5 method). These images will now be a part of your *Media Library* (http://www.podomatic.com/media/list).

TIP

Although including an image with each podcast episode is not required, the images will make your podcast episodes and podcast webpage look more interesting to users.

9. Navigate back to *Podcast Settings* (http://www.podomatic.com/settings/podcast). Enter a title, tagline, description, and tags for your podcast. Select the photo/image you uploaded to your media library for your cover art. Be sure to select a genre since this very important for landing in the right place on iTunes.

Phew! Your podcast is now set up. Thankfully, you'll never have to do that again!

10. From the *Publish* menu Publish select *Podcast* > *Create* or visit https://www.podomatic.com/quick_post.

11. In the *Upload Media* window, click *Add Files* and locate your audio file on your computer hard drive.
12. Click *Start Upload* to upload the media to Podomatic.
13. Enter a title and description for your first Podcast episode.
14. Upload a new image or use an image you have already uploaded to your media library.
15. Under *Quality* keep the default of *Music* and click *Continue.*
16. Under *Publish!* click *Publish this episode.* Podomatic will continually encourage you to "Go Pro" but this costs money and I prefer to use the free version of this service so click *No thanks, just publish.*
17. Click your podcast link that shows up in the next pop-up window to listen to your podcast episode.

Your podcast's RSS feed is the magic little bit of web code that allows you to connect your podcast to the iTunes store.

18. Navigate to your podcast management page (http://www.podomatic.com/podcast).
19. Scroll down to *Promotional Tools* and copy all of the text in the *RSS Feed* area (see Figure 6.6).

Figure 6.6
Scroll down to *Promotional Tools* and copy all of the text in the *RSS Feed* area.

20. Visit Apple's Podcast Providers website: https://itunesconnect.apple.com/itc/static/login?view=5&path=%2FWebObjects%2FiTunesConnect.woa%3F
21. Login with your Apple ID.

TIP

If you don't already have an Apple ID, create one in the iTunes application. Within the iTunes Mac/PC application, select *Store > Sign In* and click *Create Apple ID.*

22. Click the + icon in the upper left corner.
23. Paste your podcast RSS feed in the URL window and click *Validate* and then *Submit* in the upper right.

The iTunes podcast approval process typically takes a few days because real human beings actually view your podcast to make sure it is appropriate and doesn't violate the Apple iTunes terms of service.

Posting and Sharing with Google Drive

Google Drive provides the opportunity to keep file sharing within a very closed network of students with whom you manually share materials as opposed to the completely public nature of SoundCloud and Podomatic. For this reason, copyright concerns are greatly diminished. I maintain folders of play-alongs and other materials such as transcribed solos for specific tunes in my Google Drive account. I can easily share these files with students without worrying about email attachments. Many public schools are converting to Google email and Drive accounts for faculty and students. This year, all students at Haverford High School where I teach were given chromebooks with access to district-managed Google Drive accounts. I use Google Drive so extensively that I can hardly remember life before Drive!

TIP

If you created a YouTube account when we were studying Chapter 5 of this text (or if you have a Gmail account) you already have a Google Drive account!

1. Visit www.drive.google.com and enter your credentials.
2. Once in the Drive interface, click the red *New* button  in the upper left corner and select *Folder*  .
3. Name the folder "Improvisation Materials."

TIP

When using Google Drive, get in the habit of making folders to keep your content organized.

4. Double-click your new folder to open it.
5. To import materials, simply drag content from your computer into this window. After uploading, the files should appear in the Drive interface.

TIP

I recommend using the Google Chrome web browser when using Google Drive. Chrome allows you to upload folders instead of just individual files.

TIP

Be sure to allow the file to completely upload before exiting out of your web browser. Larger files or folders with lots of small files can take additional upload time.

TIP

You can drag just about anything to Google Drive including audio files, text documents, MIDI files, GarageBand files, Audacity files/folders, etc.

- GarageBand files show up as a .band folder in Drive. When you download this folder to your Mac computer, it will look more normal and open properly in the GarageBand application.
- When you upload Audacity files, be sure to drag both the .aup file and the associated "_data" folder for the project. When you download both the .aup file and _data folder to your computer, the files will open properly in the Audacity application.

6. Select one of your files and click the *Share* icon .
7. Enter the email addresses for any students with whom you'd like to share this file.
8. Click *Done* to confirm the sharing settings.

TIP

Instead of sharing an individual file, try sharing an entire folder! By sharing a folder, any files you add to the shared folder will be available to students without any extra steps. In other words, share a folder once and all content in the folder is also shared even if it is added at a later date.

TIP

When you share a file you can set the access privileges for the recipient. These privileges are primarily for files that can be edited in Google Drive such as Google Documents, Sheets, Slides, and Forms.

- Can view I most commonly select *Can view* when I share a file with students since I only want them to view or download the file.
- Can edit When I want to collaborate with someone, I will select *Can edit* so the recipient can modify the files.
- Can comment When I want to keep my material intact but allow the recipient to comment on my material, I will select *Can comment*.

TIP

You may notice *Get a shareable link* in the sharing window. I explore this option in Ch. 7 when I discuss organizing materials on a website.

9. Invite your students to login to their Drive accounts and click on *Shared with me* on the left pane of the Drive interface. The files you shared should be available for the students in this section of Google Drive.
10. To download a file from Google Drive:
    a. Select the file.
    b. Click the *More Actions* icon in the top right of the screen.
    c. Select *Download*.

TIP

Alternatively, students can click the *More Actions* icon and select *Make a Copy*. This creates a copy of the shared file and stores it in the student's Drive files. Since the teacher is the owner of the original shared file, the student would no longer have access if the teacher deleted or unshared the file. If the student makes a copy of the file, he or she will now own a copy of the file and always maintain access.

iPad Connection: Posting to SoundCloud and YouTube from GarageBand for iOS

We have seen that it is relatively easy to upload audio to SoundCloud from your computer but YouTube is a slightly different story. Since YouTube is really a video service, you must have a visual element (video or still picture) included with your audio. This can be accomplished by importing any prepared audio file into video editing software such as iMovie or Windows Movie Maker. Because video editing software opens a huge can of worms beyond the scope of this text, I prefer to use the YouTube posting feature built in to GarageBand for iOS (GBiOS) on the iPad. GBiOS automatically adds a placeholder "GarageBand" icon allowing YouTube to accept your essentially audio-only material. GBiOS also offers a simple way to upload audio to SoundCloud. I detail the steps for uploading to both YouTube and SoundCloud via GBiOS in this section.

You can use this method for material created directly in GBiOS or audio created on your computer applications such as Audacity, GarageBand, Band-in-a-Box, and Note Flight.

To import audio created using computer applications into GBiOS:

1. Connect your iPad to your computer and launch iTunes
2. In iTunes, select your device and select *Apps* in the left pane.

3. Scroll down to the *File Sharing* section. Click the *GarageBand* icon in the *Apps* window.
4. Drag audio files into the *GarageBand Documents* window.
5. Launch GBiOS on your iPad.
6. Tap the + icon and and select *Create a new song*.
7. Select *Audio Recorder* for the first track.
8. Tap the *Tracks View* icon in the upper left corner.
9. Tap the *Metronome* icon to turn it off.
10. Tap the + icon in the upper right to open the *Section Length* window.
11. Tap Section A and turn ON *Automatic* Automatic. Tap the screen to exit this window.
12. Tap the *Loops* icon in the upper right corner.
13. Tap the *Audio Files* tab Apple Loops Audio Files Music.
14. Tap and drag the audio file into the *Audio Recorder* track.

Continue following the steps below to export your audio to SoundCloud and YouTube.

To export audio from GBiOS to SoundCloud and YouTube:

15. Launch GBiOS on your iPad.
16. If an existing project opens, tap *My Songs* to view all of your GBiOS projects.
17. Tap and hold on a project. The file will have a blue outline and begin jiggling.
18. Tap the *Share* icon.
19. To upload to YouTube:
    a. Tap the YouTube icon.
    b. Sign in to your YouTube/Google account and *Allow* the necessary items.
    c. Enter a title and description for your file.
    d. Set the privacy settings. I suggest *Unlisted* so students with the link can use the material but the content will not be searchable on YouTube.
20. To upload to SoundCloud:
    a. Tap the SoundCloud icon.
    b. Sign in to your SoundCloud account.
    c. Enter a title for your file.
    d. Set the visibility, permissions, and quality. I suggest *Public* for visibility, allow both for permissions, and *Medium* for quality.

iPad Connection: YouTube Capture

When you need to make a quick video recording and upload it to your YouTube account, the YouTube capture app is the way to go. This app uses your iPad camera and microphone to record new video or it can access existing content stored in your iPad Photos app. I suggest using YouTube capture to record videos where you need to demonstrate an improvisational concept.

View YouTube Capture on the Apple App Store:

https://itunes.apple.com/us/app/youtube-capture/id576941441?mt=8

1. Download and install YouTube Capture.
2. Launch the app and be sure to allow Capture to access to your iPad's camera and microphone during initial setup.
3. Tap the record button to begin a new clip and again to end the recording.
4. Continue to record new clips as needed.
5. Tap the *Edit* icon in the upper right corner. You will now see your clips in order at the bottom of the screen.
6. Tap a clip to open the Clip Editor. Tap and drag the bars on the left and right at the bottom to trim the start/end of the clip. Tap *Done* in the upper right corner after you have finished editing.
7. Tap and hold on a clip to move it or drag it up to the trash can to delete it.
8. Tap the + in the bottom right corner and select *Add existing clip* to add videos from your iPad's Photos app into your YouTube Capture project.
9. Once your video is ready to publish, tap the arrow in the upper right corner.
10. Enter a title for your video. Adjust the privacy settings (I suggest *Unlisted*). Tap the *More* icon ••• to add a description.
11. Tap *Upload* UPLOAD in the upper right corner to send the video to YouTube.

Chapter 6 Review

1. What are the best audio file formats for posting to the Internet and why?
2. This chapter discusses how to convert audio file formats using iTunes. How could you convert an audio file using Audacity? Write a detailed list of steps to convert a Note Flight WAV file to a more Internet-friendly format using Audacity instead of iTunes.
3. SoundCloud allows other users to comment on your audio files. Discuss the pros/cons of this feature. How can you delete or hide public comments in SoundCloud?
4. Discuss the benefits of using a podcasting service such as Podomatic?
5. Organize all the materials you have created thus far in this book using Google Drive. Create a new folder for each song covered in the text and drag in exported play-along audio files, PDF notation files, etc. Share one of the folders with a student or peer with *Can View* privileges.
6. Why does YouTube not allow you to upload audio-only files? What are the workarounds to upload audio content created both on your computer and iPad to YouTube?

WEB DESIGN TO ORGANIZE ONLINE MATERIAL USING WEEBLY 7

You have created a ton of material to help teach musical improvisation throughout this text. Now you need a place to organize play-along accompaniments from Band-in-a-Box and iReal Pro, Note Flight web links, notation PDF files, various audio files created with GarageBand/GBiOS/Audacity, videos and playlists from YouTube, audio posted to SoundCloud and Podomatic, and Google Drive material.

You need a website!

Web Design Overview

In the past, creating a website could be a very technically difficult endeavor. HTML code is at the heart of most websites. If you look under the hood of a website you may see something like this:

```
<table style="width: 100%; font-family: Times New Roman;">
  <tbody>
    <tr>
      <td style="vertical-align: top;"><span style="font-size: 12pt;"><br />
      </span></td>
      <td style="vertical-align: top;"><span style="font-size: 12pt;">Name: <span
style="color: #ff4500;">Michael Fein</span><br />
      Grades: 9-12<br />
      Subjects: Digital Recording 1 and 2, Jazz Ensemble<br />
      Email: <a href="&#109;&#97;&#105;&#108;&#116;&#111;&#58;&#102;&#101;
&#105;&#110;&#64;&#104;&#97;&#118;&#115;&#100;&#46;&#110;&#101;&#116;">f
ein@havsd.net </a><br />
      Phone: 610-853-5900 ext. 2102<br />
      Homeroom Number:<span style="background-color: #ffffff;"> 2019-A</span><br />
      <br />
      <img style="width: 332px; height: 175px;" src="http://www.haverford.k12.pa.us/
cms/lib5/PA01001043/Centricity/Domain/520/fein1.jpg" alt="Fein 1" title="Fein 1"
align="" border="0" /><br />
```

```
        <img style="width: 332px; height: 569px;" src="http://www.haverford.k12.pa.us/
    cms/lib5/PA01001043/Centricity/Domain/520/fein2.JPG" alt="Fein 2" title="Fein 2"
    align="" border="0" /><br />
        </span> </td>
      </tr>
    </tbody>
</table>
<div style="font-family: Times New Roman;"></div>
```


All of this code represents a very simple page (see Figure 7.1):

Figure 7.1
Web site display of HTML code.

As you can see, HTML code is fairly complex and it takes a lot of time to learn and write. Thankfully, WYSIWYG-style (What You See Is What You Get) web design software such as Adobe's Dreamweaver allows the user enter text, images, links, etc., using a more approachable interface and the software generates the appropriate HTML code. I still recommend learning the basics of HTML code so you can dig into the code if/when the website is not displaying as you had hoped.

TIP

You can even save a Microsoft Word document as HTML for web publishing. Open any Word file and select *File > Save as Web Page*.

Next, you need a domain name, essentially your address on the web. A domain is the "www.your_website.com" address. Domain names are very inexpensive starting at about $10/year. You can purchase a .com, .net, .org, or one of the other website extensions. Some popular domain name registrars include GoDaddy(www.godaddy.com), Domain (www.domain.com), Register (www.register.com), or even Google Domains (www.domains.google). Visit one of these sites to check the availability of the domain name you'd like to purchase.

TIP

I strongly suggest purchasing "your_name.com" from one of the domain registrars listed above. Even if you don't want a website today, you may in the future and there is no easier way to find YOU on the web than YOUR_NAME.com. When I went to purchase my first domain name, www.michaelfein.com was taken so I went with www.feinmusic.com. I continued to check on the availability of www.michaelfein.com and one day it was available. I purchased it immediately and set my payment to auto-renew!

TIP

Although you can usually purchase a .net domain for less money, I always recommend going with a .com address simply because it is the most common domain extension and easiest for people to remember.

Finally, you need a hosting service. The host is where you upload the completed files you created in the web design software. Unlike domain names, hosting services typically cost significantly more per year. Whereas you may spend $10 a year for a domain name, hosting could cost $50 to $100 a year or more. The cost is higher because your actual website files are taking up space on a host's server. The host also makes your files available when other users around the world visit your website. Some popular hosting services include iPower (www.ipower.com), Host Gator (www.hostgator.com), and Dream Host (www.dreamhost.com).

TIP

Some domain registrars offer hosting services as well. Although this may seem convenient at the time, webmasters have advised me to avoid tangling these relationships. Just as when you mix business with friendships, breaking a hosting relationship with the same company that registers your domain can get sticky. It is best to select one company for domain registration and a separate company for hosting services.

Here is a summary of the entire website process:

- Create a website using WYSIWYG software such as Dreamweaver.
- Purchase a domain name for about $1 per year.
- Purchase space with a hosting service for about $50 to $100 a year and upload your website files.

Web Design with Weebly

[▶] *See Example 7.1 on the companion website for a link to an example Weebly website for this activity.*

Weebly is a completely web-based alternative to the website process listed above. I have found Weebly very useful for my students and other fellow educators. This free web design service allows a user to create multiple pages with text, photos, embedded media, and uploaded files.

The user-friendly interface is my number one reason for using Weebly. The Weebly design interface is much simpler than Dreamweaver yet the built-in templates allow the user to create a beautiful looking website. As you will see below, the user drags in blocks for all website content and builds the site with those blocks. For example, I could drag in a text block to enter text, a picture block to add a photo, and an embed block to embed web content from YouTube.

Weebly also streamlines the entire website process. Users design the website using Weebly's interface and click "publish." Weebly hosts your completed website files and generates a free domain name for each site you create ending with ".weebly.com."

TIP

Although I generally avoid tangling design, hosting, and domain registration services, Weebly is an exception because it is free and incredibly user-friendly.

TIP

Users can create multiple sites under the same account. For example, you can create a website for your improvisation materials and a separate website for something else all together and then manage both under the same Weebly account. Both websites will have their own unique domain name.

To manage multiple sites:

- Login to your Weebly account. You should see the Weebly Dashboard when you first login. You can also visit https://www.weebly.com/home/.
- Click the *Down Arrow* icon next to your first website.
- In the *My Sites* window, select *Add Site* to create a new website on your account.
- To access multiple websites in the future, click the *Down Arrow* icon and select the desired website.

TIP

If you want to avoid the ".weebly.com" ending and just have "your_name.com," you'll need to pay for the registration of that unique domain name.

Setting Up a Weebly Website

To start with Weebly, you need to create a new account.

1. Visit www.weebly.com and click *Try It Free* .
2. Enter your name, email, and password to create a free account.

TIP

Alternatively, you can sign up with your Facebook or Google Plus account.

You can begin setting up your new website.

3. Select a theme from the list and click *Choose*.
4. Weebly will prompt you to select a domain name for your new website. Enter a name into the box under *Use a Subdomain of Weebly.com* to check the availability of your desired domain name (see Figure 7.2). As previously mentioned you can also register a new domain or link a domain name you already own to avoid the ".weebly.com" domain ending.

Figure 7.2 Enter a name into the box under *Use a Subdomain of Weebly.com* to check the availability of your desired domain name.

5. After you have made your selection click *Continue.*

You are now in the Weebly design interface viewing your home page. The homepage is the first landing spot for visitors when they visit your site. It will typically contain a graphic, title header, and text welcoming the visitor.

TIP

By selecting a theme, you will have various blocks and pages already created. I suggest starting your website by modifying the existing placeholder blocks.

6. Click on a block that you'd like to modify such as the title header. If your theme does not include a *Title* block, drag one in from the left pane.
7. Replace the existing text with a title more appropriate to your website such as "Improvisation Materials."
8. If your theme does not include a graphic, drag in an *Image* block from the left pane and click it. If your theme includes a graphic, click the placeholder graphic and select *Replace Image.*
9. Drag a photo from your computer into the upload window.
10. The photo should now appear in your web page. Click the image to adjust the photo's alignment (typically centered is best) and resize as necessary.
11. Drag in a *Text* block from the left pane and click *Click here to edit.*
12. Type your welcome text into this block. Adjust formatting as needed including size, color, and alignment (typically left is best for text blocks). I suggest keeping this text short and sweet. Include a brief summary of the purpose of your website and guide the visitors to the additional pages of material linked at the top of the website.

TIP

Avoid underlining text. On the web, underlined text is almost exclusively used for links to other websites or web files. Use italics, boldface, larger font size, or different color text for emphasis.

TIP

Some themes include more blocks than you'd like to include on your homepage. Delete any unwanted blocks.

- Hover over the unwanted block with your mouse.
- Click the X in the upper right corner of the block and click the red *Delete* button to confirm (see Figure 7.3).

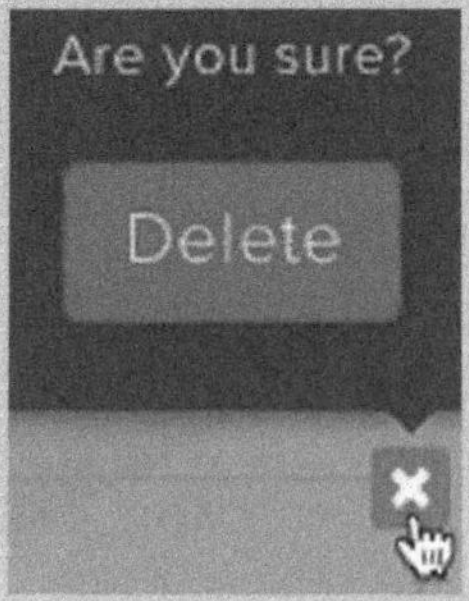

Figure 7.3
Click the X in the upper right corner of the block and click the red *Delete* button to confirm (see Figure 7.3).

13. Click *Pages* PAGES at the top of the screen. On the left pane you will see the various pages currently in your website. The first page should be called *Home*.
14. Click the second page in this list.
15. Under *Page Name*, enter a more descriptive name for the page such as "So What Materials."
16. Under *Header Type* set the style you'd prefer for this page. I prefer *Short Header* or *No Header* on non-homepages.
17. Click the left facing arrow to return to the previous screen.

TIP

Under *Header Type* click *Advanced*. Explore the various advanced fields including *Page Title* and *Page Description*.

Adding Content to a Weebly Webpage

At this point you can begin adding a significant amount of content to your webpage. You will add this content in a variety of ways including linking, uploading, and embedding. Each of these has its place in web design.

- **Link:** Use a link to create clickable text that sends users to a different website. For example, I may create a link to Miles Davis's webpage so students can learn more about this jazz legend by reading his biography. Links will almost always appear as underlined text. I strongly suggest having links open in a *new window* so users don't have to leave your website to explore linked material.
- **Upload:** Upload files for students to download. Remember that uploaded files take up space on your Weebly account. Avoid uploading large audio files and instead use the upload option for smaller documents such as notation, MIDI, or PDF files.
- **Embedding:** This is my favorite way of including web content that blends the concept of linking and uploading. The user posts material to a web service such as YouTube, SoundCloud, Podomatic, Note Flight, or Google Drive. Those services generate HTML code that the user copies and pastes into an embed code block in Weebly. Magically, the material appears in your website without taking up room on Weebly's server.

You will begin by linking text to an external website and a Google Drive folder.

1. In Weebly, click *Build* BUILD at the top of the screen.
2. Drag in a *Text* block. You will use this block to link visitors to another website.
3. Type appropriate text for a link such as "Click here to visit Miles Davis's official site."
4. Highlight the text and click the *Link* icon.
5. Click *Website URL* Website URL.
6. Enter the URL for the website you'd like to link to. Be sure to check the box for *Open link in a new window*. Click *Save* to exit this window.

TIP

Even a small typo can render your link invalid. I suggest opening the website you'd like to link to in your web browser. Then copy/paste the website URL from your web browser. This is especially important for long URLs.

7. In a new browser window, login to your Google Drive account.
   a. NOTE: In Chapter 6, I suggested uploading media to Drive and organizing this material in folders. For this example, create a new Drive folder that includes files for the tune "So What" such as a PDF of the chord changes and a play-along audio file generated by Band-in-a-Box.
8. Select a folder in Google Drive and click the *Share* icon.
9. Click *Get Shareable Link* Get shareable link in the *Share with Others* window (see Figure 7.4).
10. The default sharing settings should be *Anyone with a link can view*. Click *Copy Link* (see Figure 7.4). NOTE: Be sure to click *Done* to close the window and make your sharing settings active.

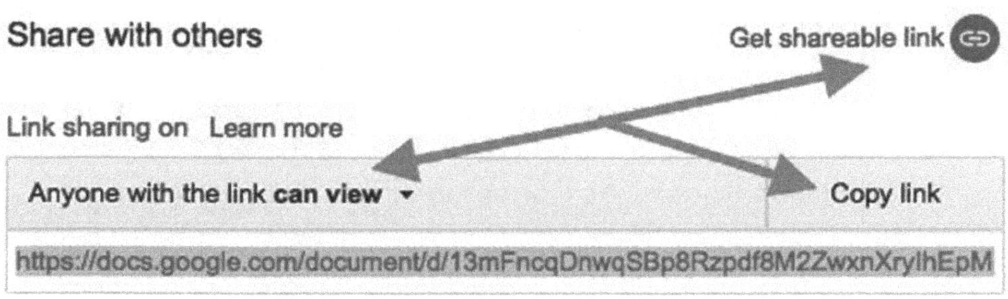

Figure 7.4
In the Google Drive *Share* window, click *Get Shareable Link* and set the sharing preferences to *Anyone with a link can view* and click *Copy Link*.

11. In Weebly, drag in another new *Text* block.
12. Type appropriate text for a link to your Google Drive folder such as "Click here to explore my *So What* improvisational materials folder."
13. Highlight the text and click the *Link* icon.
14. Click *Website URL* Website URL.
15. Paste the link into the new window under *Website URL*. Be sure to check the box for *Open link in a new window*. Click *Save* to exit this window.

TIP

Although I prefer embedding YouTube content, you can also LINK to a YouTube video. Open any YouTube video and click the *Share* icon below the video. Copy and paste the link as described above.

Next, you will upload a MIDI and PDF file to Weebly using the *File* block.

16. Open one of the Band-in-a-Box accompaniments you created in Chapter 2. I suggest opening Miles Davis's "So What."
17. In Band-in-a-Box, select *File > Save Song as MIDI File*. Select *File on Disk* and save the MIDI file to a memorable location.
18. In Weebly, drag in a *File* block.
19. Click the text *Click here to upload a file* and then click *Upload File* Upload File in the pop-up window.
20. Select the MIDI file you exported out of Band-in-a-Box. Your file is now posted to your Weebly website.
21. Login to your Note Flight account using the Google Chrome web browser.
22. Create a new score and select *Start from a Blank Score Sheet*.
23. Click the *Title* and enter "So What Chords."
24. Click the *Composer* and enter "Miles Davis, entered by *Your_Name*."
25. Double-click the bass clef staff and tap *Delete* on your keyboard to remove the staff. You should be left with a single treble clef staff.
26. Click the last bar and select *Action > Measure > Add Measure After* + Add Measure After.
27. Continue to click *Add Measure After* until you have 8 bars.

28. Click bar 1 and tap letter K on your keyboard to enter chord mode. Type "Dmin7."
29. Select bars 1–8.
30. Tap letter R three times on your keyboard to repeat the phrase 3 times.
31. Double-click the chord in bar 17 and change it to "E♭min7."
32. Click the bar line at bar 9 and tap *Return* on your keyboard. Repeat this at bars 17 and 25 (see Figure 7.5).

Figure 7.5

Chord progression for Miles Davis's "So What" entered into Note Flight.

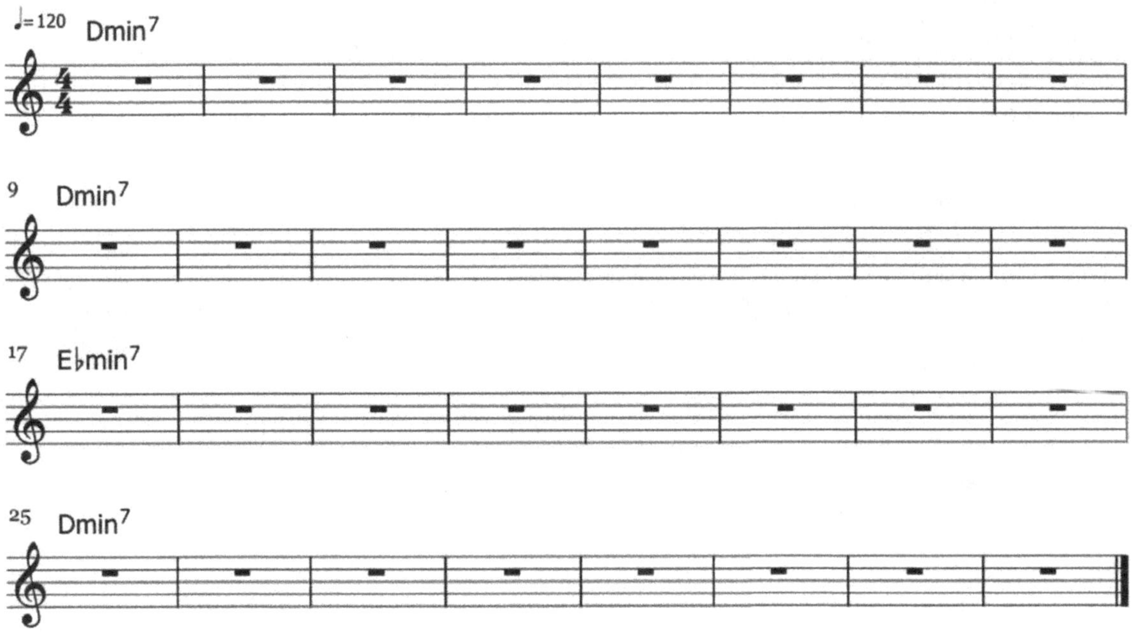

33. Type Command+P (Mac) or CTRL+P (Windows) and then click *Print* in the pop-up window.
34. In the Google Chrome print dialogue box, set your *Destination* to *Save as PDF.*
35. Name the file "So What Chords.pdf" and save it to a memorable location.
36. In Weebly, drag in another *File* block.
37. Click the text *Click here to upload a file* and then click *Upload File* in the pop-up window.
38. Select the PDF file you exported out of Note Flight.

TIP

If you have a score with multiple staves, you can still print individual parts even with a free membership to Note Flight. Click *Perform* at the top of the Note Flight interface. Set *Perform* to the single instrument staff that you'd like to save as a PDF and set *Show* to *Your Instrument*. Type Command+P (Mac) or CTRL+P (Windows). In the Google Chrome print dialogue box, set your *Destination* to *Save as PDF*. Name the file and save it to a memorable location.

TIP

You could avoid uploading files to Weebly all together by adding the files to Google Drive and creating a shareable link as described previously.

Finally, you will embed lots of web content including a YouTube video and playlist, a Podomatic episode and feed, a SoundCloud episode, and a Note Flight notation file. The process for embedding content is almost identical for every service mentioned. Essentially, you locate the embed code from one of the services, copy the embed code, and paste the code into an *Embed Code* block in Weebly. Poof! The content automatically appears!

Embedding from YouTube: YouTube

1. Sign in to your YouTube account.
2. Open a YouTube video that you'd like to embed.
3. Click *Share* under the video (see Figure 7.6).
4. Click *Embed* (see Figure 7.6).
5. Select all of the embed code and copy it (see Figure 7.6).

Figure 7.6
In YouTube, click *Share*, select *Embed*, and copy the embed code.

6. In Weebly, drag in a new *Embed Code* block.
7. Click the text *Click to set custom HTML* and paste the embed code. When you click away from the box, the YouTube video should appear.
8. In YouTube, click the *Guide* icon ≡ and open *My Channel* My Channel.
9. Click *Created Playlists* to open all of your playlists.
10. Click the name plate of one of your playlists to open the *Playlist Editor*.
11. Click the *Share* icon Share under the playlist name at the top.
12. Click *Embed* Embed.
13. Select all of the embed code and copy it.
14. In Weebly, drag in a new *Embed Code* block.
15. Click the text *Click to set custom HTML* and paste the embed code. When you click away from the box, the YouTube playlist should appear.

TIP

You will know that you embedded a playlist and not a single video because there will be a *Playlist* icon in the upper left corner of the embedded video.

TIP

You can also embed a playlist created by another YouTube user. Perform a YouTube search for a specific song, artist, or album. Any results that indicate the number of videos included are playlists (see Figure 7.7). Open one of the YouTube playlists. Click *Share* under the video and click *Embed*. The embed code here is for the entire playlist and not the single video.

Figure 7.7
Any YouTube search results that indicate the number of videos included are playlists.

Embedding from Podomatic:

1. Login to your Podomatic account (http://www.podomatic.com/podcast).
2. Under *Promotional Tools,* click *Share and Embed* (http://www.podomatic.com/badge/).
3. Click the tab for *Multi-Track Player* (see Figure 7.8).
4. Be sure your podcast is selected under *Choose a podcast or playlist* (see Figure 7.8).
5. Customize the color and size of the player (see Figure 7.8).

TIP

I suggest always selecting *No* for *Plays Automatically.* Web visitors typically find it very annoying when media automatically begins playing.

Figure 7.8
Podomatic multi-track player embed code window.

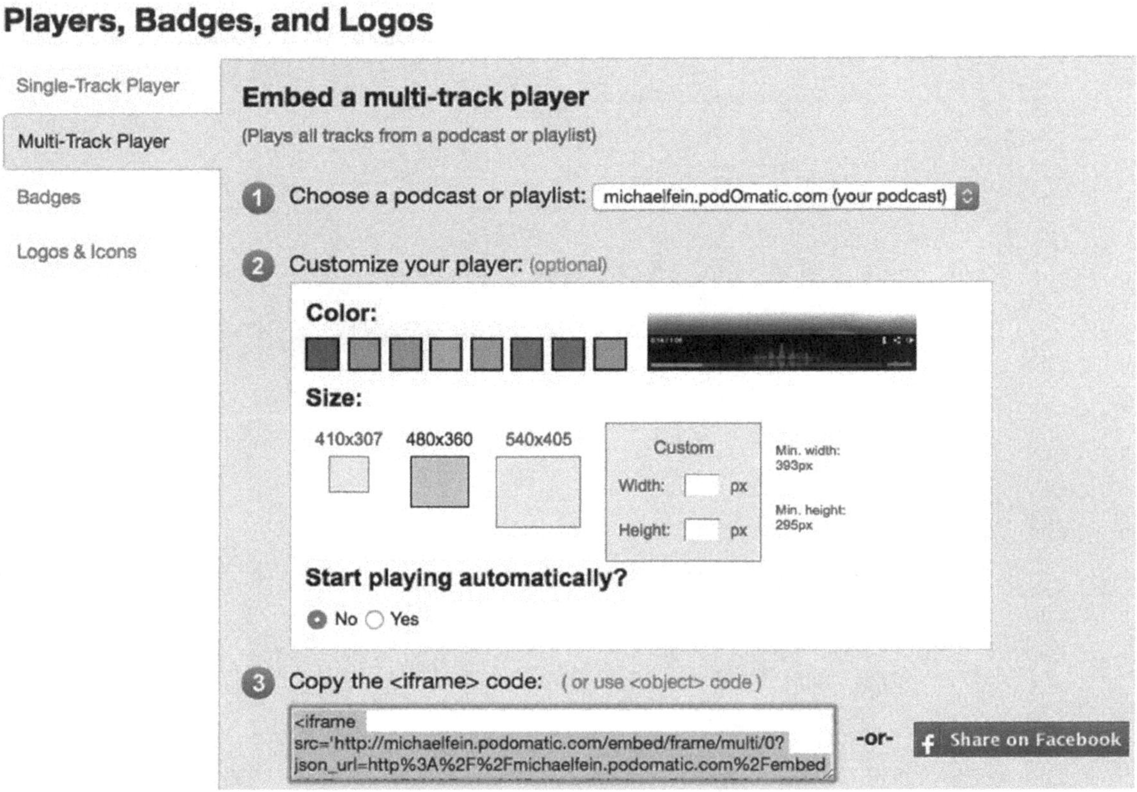

6. Select all of the embed code and copy it (see Figure 7.8).
7. In Weebly, drag in a new *Embed Code* block .
8. Click the text *Click to set custom HTML* and paste the embed code. This embeds your entire podcast feed with the latest episode ready to play. NOTE: The embedded podcast may not display correctly in the Weebly web design interface but it should work fine after you publish the website.
9. In Podomatic, still in *Share and Embed,* click the tab for *Single-Track Player* (see Figure 7.9).
10. Next to *Choose an episode* select one of your podcast episodes (see Figure 7.9).
11. Customize the color and size of the player (see Figure 7.9).
12. Select all of the embed code and copy it (see Figure 7.9).

Figure 7.9
Podomatic single track player embed code window.

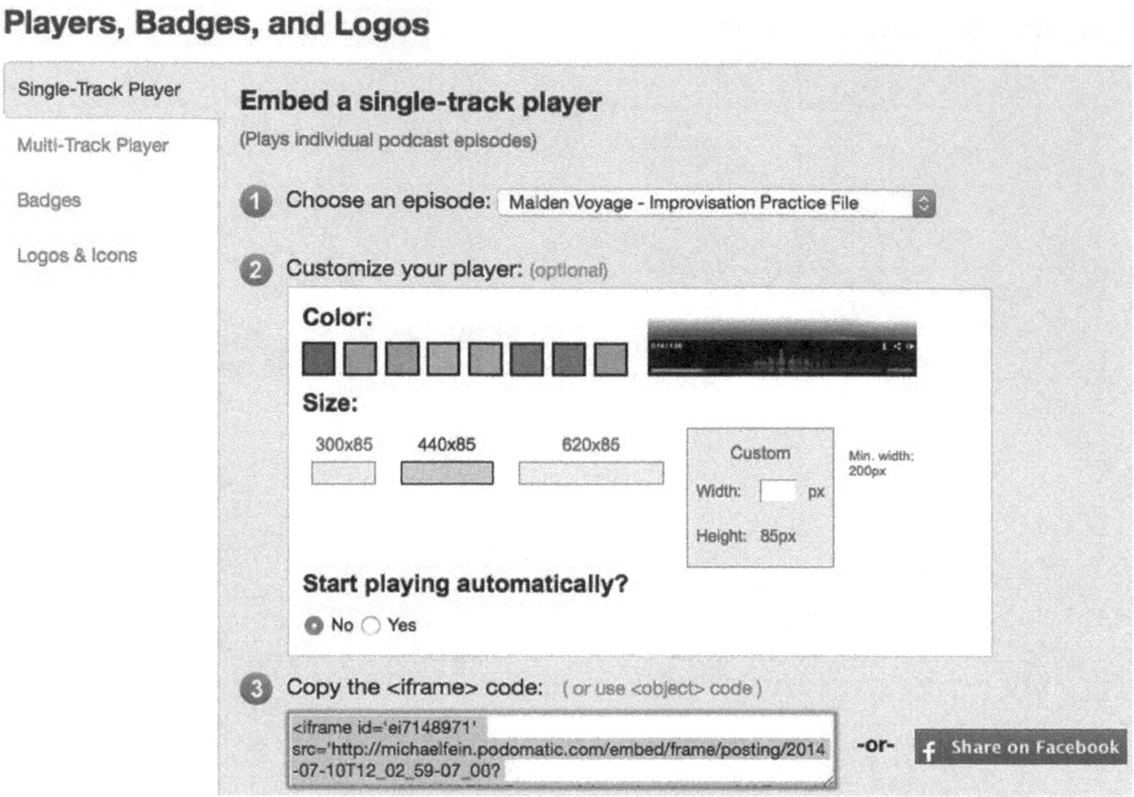

13. In Weebly, drag in a new *Embed Code* block.
14. Click the text *Click to set custom HTML* and paste the embed code. This embeds a single episode. NOTE: The embedded podcast may not display correctly in the Weebly web design interface but it should work fine after you publish the website.

TIP

If you want to have all of your podcast episodes available to your website visitors, embed the multi-track player. If you want to organize individual episodes, use the single-track player. For example, I may create a slow tempo and fast tempo version of a playlong. On my website I would embed two individual *single-track players* so I can more carefully control the order of media on my website.

TIP

You can also embed another Podomatic user's podcast feed. In Podomatic, click *Podcasts* and select one of the categories such as *Music > Jazz*. Click on one of the podcasts to open it.

- On the podcast homepage, click *Share* [Share] and copy/paste the embed code to create a multi-track player.
- Click an individual episode. On the episode page, click *Share* [Share] and copy/paste the embed code to create a single-track player.

TIP

Sometimes you will paste an embed code and the content will not appear in your Weebly web design interface. After you publish the site, the embedding will usually work just fine even if it doesn't display correctly while you are editing the site. If the content still does not embed correctly after publishing, try copying/pasting the embed code again. Even having one character incorrect can foul up the embed process.

Embedding from SoundCloud:

1. Login to your SoundCloud account. When you login, you should be at your account stream.
2. Click the *Share* icon below any of your audio posts.
3. Click *Embed* (see Figure 7.10).
4. Select one of the two SoundCloud embedded player styles. I personally prefer the skinnier player because these take up less space when stacked on top of each other on a website (see Figure 7.10).
5. Select all of the embed code and copy it (see Figure 7.10).

Figure 7.10 SoundCloud embed code window.

6. In Weebly, drag in a new *Embed Code* block.
7. Click the text *Click to set custom HTML* and paste the embed code. When you click away from the box, the SoundCloud audio player should appear.

TIP

You can also embed other SoundCloud users' audio posts. Perform a search on the SoundCloud website. Click *Share* under any search result. Click *Embed* and copy the embed code.

Embedding Note Flight notation:

1. Login to your Note Flight account and open an existing score.
2. Click *Connect* in the upper right corner.
3. In the *Connect* window, click *Change* next to *Only you can access this score* (see Figure 7.11).

4. Check the box for *Everyone* and set *Anyone with a link can VIEW* (see Figure 7.11).
5. Click the *Embed* icon </> (see Figure 7.11).
6. Select all of the embed code and copy it (see Figure 7.11).
7. **Click *Save* before leaving this page.** If you don't click *Save* your settings won't be saved and the embed process will not work (see Figure 7.11).

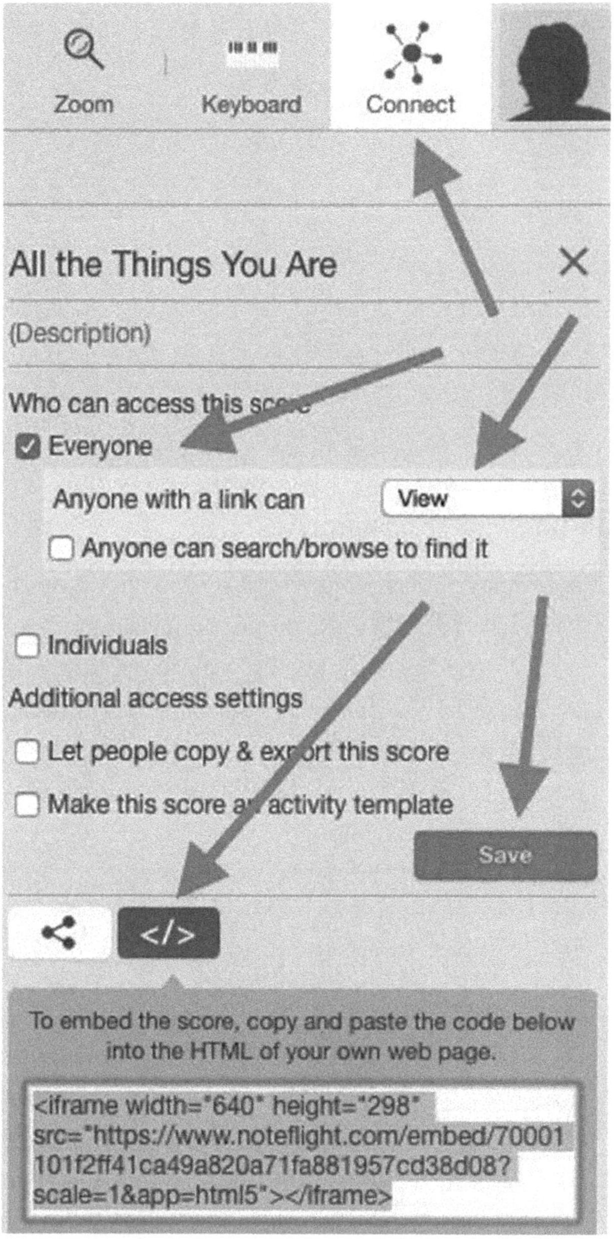

Figure 7.11
Note Flight embed code window.

8. In Weebly, drag in a new *Embed Code* block.
9. Click the text *Click to set custom HTML* and paste the embed code. You should now see the actual notation from Note Flight. You will also have playback controls (see Figure 7.12).

Figure 7.12
Embedded Note Flight file window.

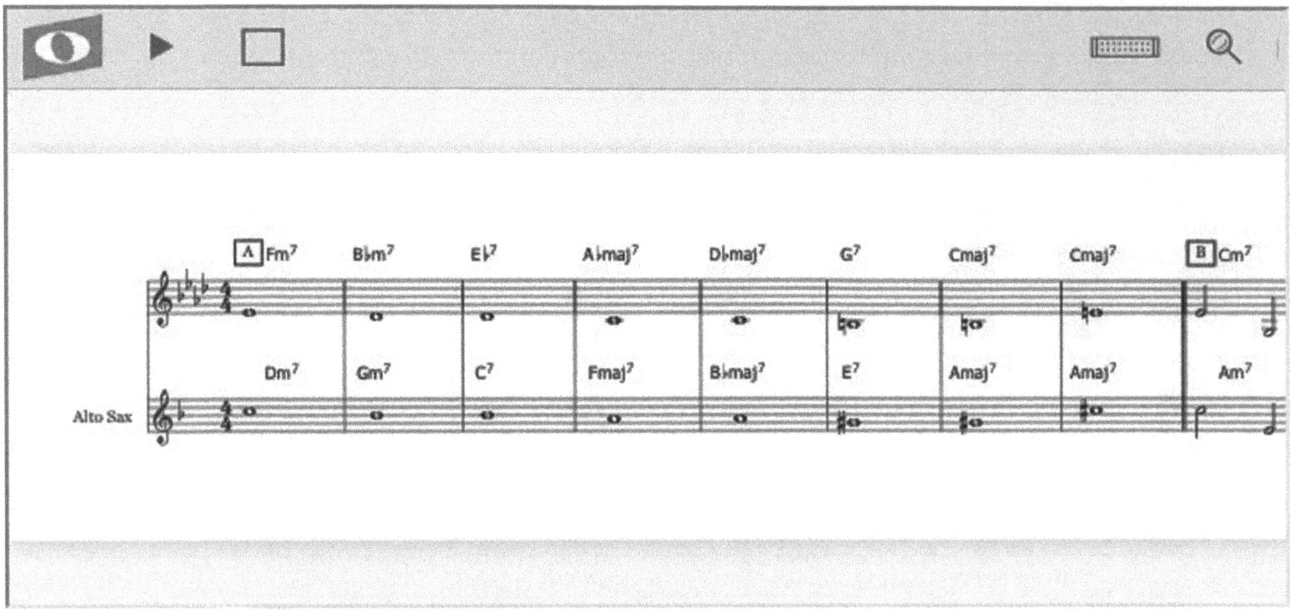

Publishing Your Website

It is time to publish your website so it is LIVE on the Internet. Simply click the large orange *Publish* icon PUBLISH in the upper right corner of the Weebly web design interface.

Click the link (www.your_website.weebly.com) to view the live page on the Internet (see Figure 7.13). This will give you an idea of what visitors will see. Be sure to troubleshoot anything that isn't working as you intended. As you continue to develop your website with additional content/pages you will need to publish each time for the changes to take effect.

Figure 7.13
After publishing your Weebly website, click the link to view the live page on the Internet.

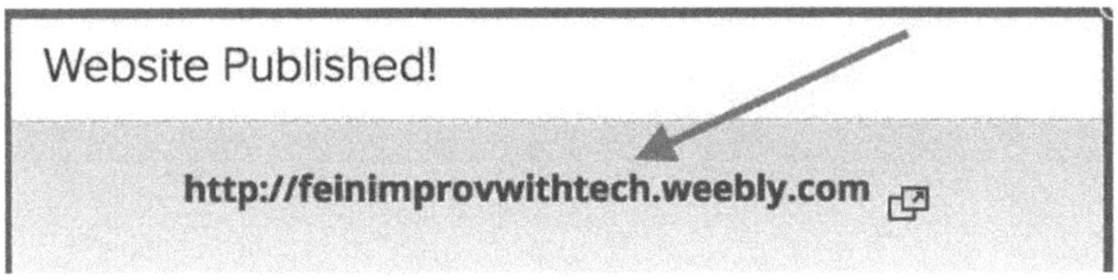

TIP

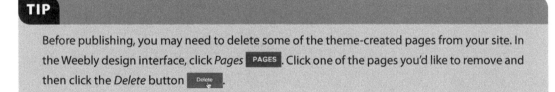

Before publishing, you may need to delete some of the theme-created pages from your site. In the Weebly design interface, click *Pages* PAGES. Click one of the pages you'd like to remove and then click the *Delete* button Delete.

TIP

Developing a website takes time. Think of your website as an organism that changes, grows, and develops over time. Continue to add new pages with additional content to your site.

TIP

Try viewing your website on a smartphone or tablet. Weebly typically does a nice job of making your website mobile-compliant. In the Weebly design interface click the *Desktop* icon and select *Mobile* Mobile to preview your site on a mobile platform. I still recommend actually viewing the site on a mobile device to confirm the functionality of your website.

NOTE: Apple iOS devices will not display Flash content and therefore some embedded material (such as Flash-based Note Flight embedded scores) will not display on an iPhone or iPad. Consider purchasing a third-party app from the Apple App Store such as Puffin Browser Pro ($3.99) or Photon Browser ($4.99).

Puffin Browser Pro on App Store:

- iPhone/iPad ($3.99): https://itunes.apple.com/us/app/puffin-browser-pro/id406239138?mt=8

Photon Browser on App Store:

- iPad ($4.99): https://itunes.apple.com/us/app/photon-flash-player-for-ipad/id430200224?mt=8
- iPhone ($3.99): https://itunes.apple.com/us/app/photon-flash-player-for-iphone/id453546382?mt=8

iPad Connection: Weebly App

The Weebly app allows you to build or edit your website on a mobile platform. With the exception of YouTube content, you cannot embed other web content using the app. You can, however, do most other common web design tasks including adding text, images, and even files. In the steps below, I will guide you through building a new page on your current Weebly website all through the Weebly iPad app.

Download the Weebly App (iPhone/iPad) from the Apple App Store:

https://itunes.apple.com/us/app/weebly-create-free-website/id511158309?mt=8

1. Launch the Weebly app on your iPad.
2. Tap *Log In* and enter your credentials.
3. Tap *Sites* Sites in the upper left corner to access your various sites managed under your account.

TIP

In the *Sites* menu, tap the + icon to create a new site directly from the iPad app.

4. Tap *Edit Site* Edit Site to enter the familiar Weebly design interface.
5. Tap *Pages* in the top right.
6. Click the + icon to add a new page.

7. Enter a name for the page under *Page Name* such as "Little Sunflower Materials" and tap *Save* in the upper right corner of this window.
8. Tap *Build* in the top right.
9. Tap and drag in a *Title* block.
10. Enter a title for this page such as "Little Sunflower Materials." Tap anywhere on the screen to exit out of the *Title* block editing.
11. Tap *Build* and drag in a *Text* block.
12. Enter appropriate text about the tune, "Little Sunflower," such as, "Little Sunflower is a modal tune based on 3 different scales (D dorian, E♭ ionian, and D ionian)."

You will now create a custom image using Note Flight and the iPad Photos app.

13. Launch Safari on your iPad.
14. Visit www.noteflight.com and login to your account.
15. Tap *New Score*. You will now enter the familiar Note Flight interface.
16. Tap (*Title*) and enter "Little Sunflower."
17. Tap (*Composer*) and enter "Freddie Hubbard."
18. Tap the lines and spaces of bar 1 in the treble clef staff.
19. Using the on-screen piano keyboard enter a single D note.
20. Tap *Action* > *Duration* > *Eighth Note* to change the the rhythm to an eighth note.
21. Using the on-screen piano keyboard enter the notes to the D dorian, E♭ major, and D major scales (see Figure 7.14).

TIP

To change the enharmonic spelling of a note:

- Tap the note.
- Tap *Action* > *Pitch* > *Enharmonic Shift* Enharmonic Shift.

Figure 7.14

Using the on-screen piano keyboard in the Note Flight iPad interface, enter the notes to the D dorian, E♭ major, and D major scales.

22. Tap bar 4, tap the rectangle above the bar, and then tap the "-" icon to delete the bar.
23. Tap the rectangle to the left of the treble staff to select all three measures (see Figure 7.15).

Figure 7.15

Tap the rectangle to the left of the treble staff to select all three measures.

24. Tap *Action* > *Edit* > *Copy* [Copy].
25. Tap the lines and spaces of bar 1 of the bass clef staff.
26. Tap *Action* > *Edit* > *Paste* [Paste].
27. Tap *Action* > *Pitch* > *Move Down an Octave* [Move Down an Octave].
28. Tap the first note in bar 1 of the treble clef staff.
29. Tap *Action* > *Text* > *System Text* [System Text] and type "D dorian."
30. Tap the first note in bar 2 of the treble clef staff.
31. Tap *Action* > *Text* > *System Text* [System Text] and type "Eb major."
32. Tap the first note in bar 3 of the treble clef staff.
33. Tap *Action* > *Text* > *System Text* [System Text] and type "D major" (see Figure 7.16).

Figure 7.16

Type the scale names above each measure using *System Text*.

34. Take a screenshot by pressing the iPad's home button and power button at the same time.
35. Tap the iPad's home button to exit out of the Safari app.
36. Launch the Photos app. This app is included with iOS.
37. Scroll through your photos to locate the screenshot you captured in the previous step.
38. Tap the screenshot to open it in full screen view.
39. Tap *Edit* in the upper right corner.
40. Tap the *Crop* icon [Crop icon] and adjust the image so only the notated scales are visible (see Figure 7.17).

Figure 7.17
Crop the screenshot image using the iPad Photos app.

41. Tap *Done* in the upper right corner to save the image edit.
42. Tap the iPad's home button to exit the Photos app.
43. Launch the Weebly app again.
44. Tap *Build* and drag in an *Image* block. Be sure to allow the Weebly app access to your camera roll.
45. Tap *All Photos*, locate the screenshot you cropped in the previous steps, and tap it to select the image.
46. Tap *Done* to add the image.

TIP

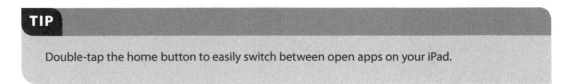

Double-tap the home button to easily switch between open apps on your iPad.

You will now embed YouTube content in your website.

47. Launch the YouTube app on your iPad.

48. Open a video you'd like to embed on your Weebly page such as a recording of Freddie Hubbard performing "Little Sunflower."
49. Tap the video and tap the *Share* icon.
50. Tap *Copy Link.*
51. Tap the iPad's home button to exit the YouTube app.
52. Launch the Weebly app.
53. Tap *Build* and drag in a *YouTube* block.
54. Under *YouTube Video URL*, double tap and select *Paste.*
55. Tap anywhere on the screen to exit and your YouTube content should now be embedded.

TIP

Since the Weebly iPad app allows you to embed YouTube content only, create audio content in the GarageBand for iOS app and share it to YouTube. Copy/paste the YouTube URL for this content into a Weebly *YouTube* block.

Finally, you will upload an audio file from iReal Pro to Google Drive.

NOTE: You must have the Google Drive app installed to complete this process.

56. Launch iReal Pro.
57. Open an existing accompaniment or create a new one from scratch. I suggest opening or creating an accompaniment for "Little Sunflower." Remember, you can also search the iReal Pro Forum for tunes entered by other users.
58. Tap the *Share* icon in the upper right corner.
59. Tap *Share Audio* Share Audio and select *AAC* AAC . iReal Pro will now export the audio file.
60. After exporting is complete, swipe to the left in the first row of icons and tap the *More* icon.
61. Turn on sharing to Google Drive and tap *Done.* Google Drive
62. Tap the Google Drive icon.
63. Tap *Upload* UPLOAD in the pop-up window.
64. Tap *Close* in the upper left corner of this window to exit the *Upload* window.

TIP

If you have multiple Google accounts registered on your iPad, be sure to take note of which account you are using to save your audio file. The Google Drive account is listed in the bottom of the *Upload* window.

65. Tap the iPad's home button to exit iReal Pro.
66. Launch the Weebly app once more.
67. Tap *Build* and drag in a *File* block.

68. Tap *Locations* in the upper left corner of the pop-up window and select Google Drive (see Figure 7.18).

Figure 7.18
Add a file into a Weebly *File* block by tapping *Locations* in the upper left corner of the pop-up window and selecting Google Drive.

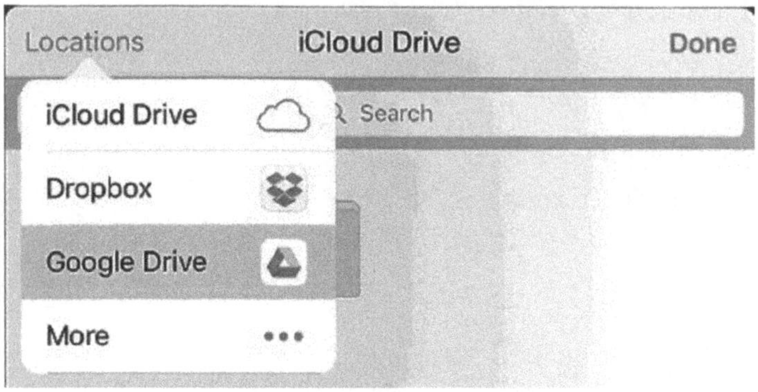

TIP

If Google Drive is not an option under *Locations*, tap *More* and turn on sharing with Google Drive.

69. Locate the iReal Pro audio file you exported previously and select it. Weebly will upload the file to your account and create a download link.

TIP

Instead of scrolling through all of your Google Drive files, tap the *Search* field and enter the name of the iReal Pro file to locate it quickly.

TIP

You can also add any material to your Google Drive account from your desktop/laptop computer and access these files via the Weebly and Google Drive iPad applications.

70. Tap *Preview* Preview in the upper right corner of the Weebly app.
71. Peruse your website. If you are happy with everything, tap *Publish* Publish to get your new webpage live on the Internet.
72. Tap *Done* when publishing is complete to exit the window or tap your website link to view your website live on the Internet (see Figure 7.19)!

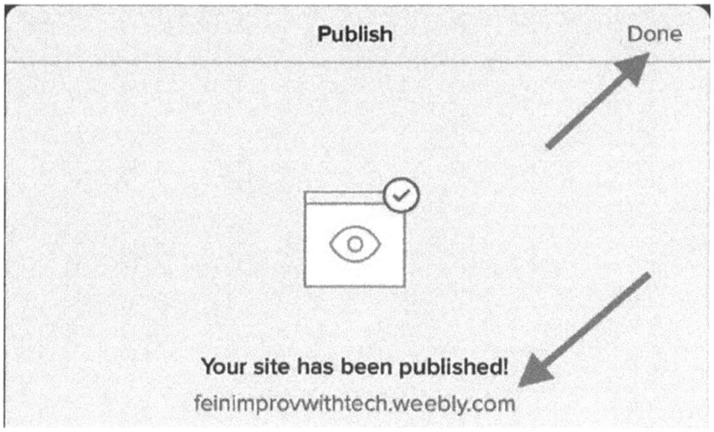

Figure 7.19
Tap *Done* when publishing is complete to exit the window or tap your website link to view your website live on the Internet.

Chapter 7 Review

1. What is HTML code and what type of software has helped web designers avoid typing so much HTML code?
2. What is a domain name and why do you need one? What is the approximate yearly cost?
3. Search a web registry site for possible domain names for your website and list five potential domain names.
4. What is the purpose of a hosting service? What is the approximate yearly cost?
5. Explain the benefits of using Weebly for a novice webmaster.
6. Explain the difference between linking, uploading, and embedding content on your website. Provide an example of how you would use each method and why you chose that method for that specific content.
7. When using the Weebly iPad app for web design, how can you incorporate media via a file upload since general embedding is not currently possible? How could you use the *YouTube* block in Weebly along with GarageBand for iOS to embed media from your iPad? How could you take an audio file from your computer and upload it to Weebly via the Weebly iPad app using Google Drive?

Closing Remarks

I hope the theory and technology activities presented in this text have served to improve both your personal musical improvisation skills and your pedagogical approach to teaching musical improvisation to students. I feel strongly that improvisation is an essential skill for all musicians at every level of development. Improvisation is a unique musical activity that combines theory, ear training, and technical skill. Improvisation allows the musician to compose in the moment and express his or her individuality. Through improvisation, students learn self-confidence and accept vulnerability on the musical stage.

As a teacher, the best way to teach improvisation is to experience improvisation for yourself. Practice the activities outlined in the book on your main instrument on your own. Discover the challenges you encounter and try to articulate these to your students. Let your students know that you too are learning and growing as an improviser.

The improvisation and music theory concepts presented in the text are time-tested; there is no risk of these becoming out-dated anytime in the foreseeable future. Just about all improvising musicians have worked through these ideas in one form or another. The technology end of things, however, is ever-changing. I know it is quite possible that some of the steps outlined in each chapter will change slightly or drastically with an application update or website revamp. This is unavoidable in the world of technology but it can be understandingly frustrating. For example, as I was writing this book, Note Flight released an entirely new version of its web interface. Did that update negate the overall goal of the Note Flight activities from Chapter 3? Absolutely not. I simply needed to invest some time navigating the new interface to figure out how to accomplish the same musical goal within the framework of the updated application. My biggest suggestion is to find the technology tools that meet your needs, stick with them, and use them as often as possible. You do not need to become an expert with every application or service presented in this book, but you do need to consistently work at being fluent in the application(s) you find most appropriate for your students.

Finally, remember that technology is a tool. Avoid letting the technology consume too much of your time, taking away from the ultimate musical goal of teaching improvisation. Think first about the musical concept you'd like to teach and then look toward the most appropriate technology tool to enhance your instruction in an efficient manner.

Please feel free to reach out to me with questions or comments via email (michael@feinmusic.com). I'd love to hear about how improvisation has impacted you and your students.

Index